BIOTECHNOLOGICAL TECHNIQUES

(PRACTICAL MANUAL)

BIOTECHNOLOGICAL TECHNIQUES

(PRACTICAL MANUAL)

By

Anupam Singh
Assistant Professor
Department of Biotechnology
S.D. College of Engineering and Technology,
MUZAFFARNAGAR-251001

and

Ashwani Kumar Singh
Senior Research Fellow
College of Biotechnology
Sardar Vallabhbhai University of Agri. and Tech.,
MEERUT-250110

NPH

NARENDRA PUBLISHING HOUSE

DELHI-110006 (INDIA)

Bio-Science Practical Series

First Edition 2011

ISBN : 978-93-80428-26-0

Published by :

NARENDRA PUBLISHING HOUSE

Publisher and Distributor

1417, Kishan Dutt Street, Maliwara,

DELHI-110006 (India)

Phones: 91-011-23268470, 91-011-23259412

E-mail: info@nphindia.com **Website:** www.nphindia.com

Laser Typeset by Amrit Graphics, Shahdara, Delhi 110 032

Printed in India

Dedicated with
profound gratitude to our
parents

CONTENTS

PREFACE

The objective of this practical manual is to give a practical knowledge to the researchers, students and to the biotechnologists in different field of life sciences. This manual is a core collection of experiments of biotech sciences, which will be very useful for those who desire to have fundamental practicals. Biotechnology has become a part of our lives. Applications of biotechnological techniques have provided us many benefits, and a better quality of life. The world today uses a language which has a lot of biotechnological science in it. Without knowing, we use many words and phrases derived from biotech science. We are also becoming conscious of our environment as well as our economy. Biotechnology has much to do with both these aspects. It is, therefore, important to learn the language of biotechnological science. Children and adults alike have to know the rudiments of science and must be able to use the language of biotech science where necessary. They must be able to apply the lessons learnt from biotech science in daily life. It is for this purpose that we have produced a practical manual entitled *"Biotechnological Techniques: Practical Manual"* in six chapters. The manual has the following six chapters:

Chapter 1. General instructions and Laboratory Methods

Chapter 2. Biochemistry

Chapter 3. Microbiology

Chapter 4. Cell Biology and Genetics

Chapter 5. Molecular Biology

Chapter 6. Tissue Culture Techniques

The first chapter, i.e., 'General instruction and laboratory methods' gives information about the general instruction about safety aspects, precaution and laboratory methods. The remaining chapters have included the related experiments to their name of chapters respectively. The practical manual has also contained the important information about the solution, buffers, media preparation and their calculation methods. It emphasizes on the practical approach. We believe that this practical manual will expose students to practical approach towards their research area.

It describes various aspects of experiments in simple language. It is hoped that this will be useful to biotech students of various universities, research institutes and UPTU students as supplementary reading material and to all others who want to learn biological techniques and partake in the excitement of this experience. We also invite the valuable suggestions and advice particularly from students to make the practical manual good enough.

Anupam Singh

Ashwani Kumar Singh

Acknowledgement

In the completion of this practical manual we are thankful to our seniors, colleagues, juniors and those who have involved directly or indirectly in this achievement. We also acknowledge the encouragement and motivation imparted by persons who inspired us to take this opportunity especially, Dr. S. N. Chauhan (Director, S.D. College of Engineering & Technology, Muzaffarnagar), Dr. A.K. Gautam (Principal, S.D. College of Engineering & Technology, Muzaffarnagar), Dr. P. Pundir (Dean, S.D. College of Engineering & Technology, Muzaffarnagar), Dr. Raghvendar (ARS, Prof. & Head, Immunology, SVPUA & T, Meerut), Dr. Pramod Kumar Rai (Principal Scientist, ICAR), Dr. Ravindra Kumar (Assoc. Prof., Dept. of Cell Biology, SVPUA & T, Meerut), Dr. Pankaj Kumar (Assoc. Prof., Dept. of Biochemistry & Physiology, SVPUA & T, Meerut), Dr. Akash Tomar (Asst. Prof., Dept. of Recombinant Technology, SVPUA & T, Meerut), Dr. Sangita Bansal (Asst. Prof., Dept. of Commercial Biotechnology, SVPUA & T, Meerut) and Dr. Jitender Singh (Asst. Prof., Dept. of Immunology, SVPUA & T, Meerut), for their precious suggestions and their valuable guidance. We are also highly appreciated by the faculty members of S.D. College of Engineering & Technology, Muzaffarnagar, specially by Mr. Anoop Kumar, Mr. Naveen Dwivedi, Miss Avani agarwal, Miss Vibha shankhwar, Mrs Subha Dwivedi. We are especially thankful to Er. Prateek Gupta (Lecturer, S.D. College of Engineering & Technology, Muzaffarnagar) for his kind cooperation during completion of this manual.

ABBREVIATIONS

ATP	Adenosinetriphosphate
Å	Angstrome
cm	Centimeter
C	Carbon
^{0}C	Degree centigrade
EM	Electron microscope
gm/gms	gram
i.e.	that is
LB	Laurea broth
L/lit	Liter
mm	Millimeter
mM	Milimole
ml	Milliliter
m	Mole
mg	miligram
MW	Molecular weight
min	minute
N	Normal
nm	nanometer
NA	Numerical aperture
ng	nanogram
%	Percentage
rpm	Revolution per minute
h/hr/hrs	Hours
SEM	Scanning electron microscope

SDW	Sterile distilled water
TEM	Transmission electron microscope
Psi	Pound per second inch
λ	Lamda
μm	Micrometer
μg	Microgram
w/v	Weight/volume
v/v	Volume/volume
μl	microliter

1 General Instructions and Laboratory Methods

SAFETY PROCEDURES

(a) Chemicals : A lot of chemicals used in the laboratory are hazardous for human beings. All manufacturers of hazardous materials are required by law to supply the user with relevant information on any hazardous associated with their chemicals. This is suggested to take information which is supplied along with the chemicals. This information contains the chemical name, health hazard data, including first aid treatment, physical data, fire and explosion hazard data, reactivity data, spill or leak procedures, and any special precautions needed when handling this chemical. This is strongly urged that first be familiar with the concern chemical by the supplied information prior to use them, and certainly in the case of any accidental exposure or spill.

The following chemicals are particularly noteworthy:

Phenol = can cause severe burns

Acrylamide = potential neurotoxin

Ethidium bromide = carcinogen.

These chemicals are not harmful if used properly, always wear gloves when using potentially hazardous chemicals, and never mouth pipette them. If you accidentally splash any of these chemicals on your skin, immediately rinse the area thoroughly with water and inform the instructor. Discard waste in appropriate containers.

(b) Ultraviolet Light : Exposure to ultraviolet (UV) light can cause acute eye irritation. Since the retina cannot detect UV light, you can have serious eye damage and not realize it until 30 minutes to 24 hours after exposure. Therefore, always wear appropriate eye protection when using UV lamps.

(c) **Electricity :** The voltages used for electrophoresis are sufficient to cause electrocution. Cover the buffer reservoirs during electrophoresis. Always turn off the power supply and unplug the leads before removing a gel.

(d) **General Housekeeping :** All common areas should be neat and clean and all dirty dishes, electrophoresis equipment, etc., should be dealt with appropriately. Since you have only a limited amount of space of your own, it is to your advantage to keep that area clean. Since you will use common facilities, all solutions and everything stored in an incubator, refrigerator, etc., must be labeled properly. In order to limit confusion, each person should use his initials or another unique designation for labeling plates, etc. Unlabeled material found in the refrigerators, incubators, or freezers may be discarded. Always mark the backs of the plates with your initials, the date, and relevant experimental data, e.g., strain numbers.

PREPARATION OF SOLUTIONS

(a) Calculation of Molar, %, and "X" Solutions :

(i) A molar solution is one in which 1 liter of solution contains the number of grams equal to its molecular weight.

Example: To make up 100 ml of a 5M NaCl solution = 58.456 (mw of NaCl) g x 5 moles x 0.1 liter = 29.29 g in 100 ml sol mole liter.

(ii) Percent solution :

Percentage (*w*/*v*) = weight (g) in 100 ml of solution

Percentage (*v*/*v*) = volume (ml) in 100 ml of solution.

Example : To make a 0.7% solution of agarose in TBE buffer, weigh 0.7g of agarose and bring up the volume to 100 ml with the TBE buffer.

(iii) "X" solutions. Many enzyme buffers are prepared as concentrated solutions:

Example : 5X or 10X (5 or 10 times the concentration of the working solution), and are then diluted so that the final concentration of the buffer in the reaction is 1 X.

Example : To set up a restriction digestion in 25 ml, one would add 2.5 ml of a 10X buffer, the other reaction components, and water for a final volume of 25 ml.

(b) Preparation of Working Solutions from Concentrated Stock Solutions : Many buffers in molecular biology require the same components, but often in varying concentrations. To avoid having to make every buffer from scratch, it is useful to prepare several concentrated stock solutions and dilute as needed.

Example. To make 100 ml of TE buffer (10 mM Tris, 1 mM EDTA), combine 1 ml of a 1 M Tris solution and 0.2 ml of 0.5 M EDTA and 98.8 ml sterile water. The following is useful for calculating amounts of stock solution needed:

$$C_i \times V_i = C_f \times V_f$$

Where C_i = initial concentration, or concentration of stock solution

Vi = initial volume, or amount of stock solution needed

C_f = final concentration, or concentration of desired solution

V_f = final volume, or volume of desired solution.

(c) Steps in Solution Preparation :

(i) Refer Appendix-B and C of this laboratory manual for any specific instructions on preparation of the particular solution. The bottle must be labeled for any specific precautions in handling the chemical.

(ii) Weigh out the desired amount of chemical (s). Use an analytical balance if the amount is less than 0.1 g.

(iii) Pour the chemical(s) in an appropriate size beaker with a stir bar.

(iv) Add less than the required amount of water. Prepare all solutions with double-distilled water (in a carboy).

(v) When the chemical is dissolved, transfer to a graduated cylinder and add the required amount of distilled water to achieve the final volume. An exception is when preparing solutions containing agar or agarose. Weigh the agar or agarose directly in the final vessel.

(vi) If the solution needs to be at a specific pH, check the pH meter with fresh buffer solutions and follow the instructions for using a pH meter.

(vii) Autoclave, if possible, at 121°C for 20 minutes. Some solutions cannot be autoclaved; for example, SDS. These should be filter-sterilized through a 0.22mm filter. Media for bacterial cultures must be autoclaved the same day it is prepared, preferably within an hour. Store at room temperature and check for contamination prior to use by holding the bottle at eye level and gently swirling it.

(viii) Solid media for bacterial plates can be prepared in advance, autoclaved and stored in a bottle. When needed, the agar can be melted in a microwave, any additional components, e.g., antibiotics, can be added, and the plates can then be poured.

(ix) Concentrated solutions, e.g., 1M Tris-HCl pH = 8.0, 5M NaCl, can be used to make working stocks by adding autoclaved double-distilled water in a sterile vessel to the appropriate amount of the concentrated solution.

(d) Glassware : Glass and plasticware used for molecular biology must be clean well. Glassware should be rinsed with distilled water and autoclaved or baked at 150°C for 1 hour. For experiments with RNA, glassware and solutions are treated with diethylpyrocarbonate to inhibit RNases, which can be resistant for autoclaving.

Plasticware, such as pipettes and culture tubes, is often supplied sterile. Tubes made of polypropylene are turbid and resistant to many chemicals, like phenol and chloroform; polycarbonate or polystyrene tubes are clear and not resistant to many chemicals. Micropipette tips and microfuge tubes should be autoclaved before use.

DISPOSAL OF BUFFERS AND CHEMICALS

(i) Any uncontaminated, solidified agar or agarose should be discarded in the trash, not in the sink, and the bottles rinsed well.

(ii) Any media that becomes contaminated should be promptly autoclaved before discarding it. Petri dishes and other biological waste should be discarded in biohazard containers, which will be autoclaved prior to disposal.

(iii) Organic reagents, e.g., phenol, should be used in a fume hood and all organic waste should be disposed of in a labeled container, not in the trash or the sink.

(iv) Ethidium bromide (EtBr) is a mutagenic substance that should be treated before disposal and handled only with gloves. Ethidium bromide should be disposed off in a labeled container.

EQUIPMENT

(a) **General Comments :** Keep the equipment in good working condition. Don't use anything (any instrument) unless you have been instructed in its proper use. Report any malfunction immediately. Rinsc out all centrifuge rotors after use, in particular if anything spills. Please do not waste supplies use only what you need. If the supply is running low, please notify the instructor before it is completely exhausted. Occasionally, it is necessary to borrow a reagent or equipment from another lab; notify the instructor.

(b) **Micropipettors :** Most of the experiments you will conduct in the laboratory will depend on your ability to accurately measure volumes of solutions using micropipettors. The accuracy of your pipetting can only be as accurate as your pipettor, and several steps should be taken to ensure that your pipettes are accurate and maintained in good working order. Then they should be checked for accuracy following the instructions given by the instructor. If they need to be recalibrated, do so. There are 2 different types of pipettors, Rainin pipetmen and Oxford benchmates. Since the pipettors will use different pipette tips, make sure that the pipette tip you are using is designed for your pipettor.

(c) **Using pH Meter Biological functions :** are very sensitive to changes in pH and hence, buffers are used to stabilize the pH. A pH meter is an instrument that measures the potential difference between a reference electrode and a glass electrode, often combined into one combination electrode. The reference electrode is often AgCl2. An accurate pH reading depends on standardization, the degree of static charge, and the temperature of the solution.

(d) **Autoclave Operating Procedures :** Place all material to be autoclaved on an autoclavable tray. All items should have indicator tape. Separate liquids from solids and autoclave separately. Make sure the lids on all bottles are loose.

Make sure the chamber pressure is at zero before opening the door.

WORKING WITH DNA

(a) **Storage :**

- The following properties of reagents and conditions are important considerations in processing and storing DNA and RNA. Heavy metals promote phosphodiester breakage. EDTA is an excellent heavy metal chelator.
- Free radicals are formed from chemical breakdown and radiation and they cause phosphodiester breakage. UV light at 260 nm causes a variety of lesions, including thymine dimers and crosslinks. Biological activity is rapidly lost. 320 nm irradiation can also cause crosslinks.
- Ethidium bromide causes photo-oxidation of DNA with visible light and molecular oxygen. Oxidation products can cause phosphodiester breakage. If no heavy metals are present, ethanol does not damage DNA.
- 5°C is one of the best temperatures for storing DNA. –20°C causes extensive single- and double-strand breaks.
- –70°C is probably excellent for long-term storage. For long-term storage of DNA, it is best to store it in high salt (>1 M) in the presence of high EDTA (>10 mM) at pH 8.5.
- Storage of DNA in buoyant CsCl with ethidium bromide in the dark at 5°C is excellent.

(b) **Purification :** To remove protein from nucleic acid solutions:

(i) Treat with proteolytic enzyme, e.g., pronase, proteinase K.

(ii) Phenol Extract. The simplest method for purifying DNA is to extract with phenol or phenol: chloroform and then chloroform. Phenol denatures proteins and the final extraction with chloroform removes traces of phenol.

(iii) Use CsCl/ethidium bromide density gradient centrifugation method.

(c) Quantitation :

(i) Spectrophotometric : For a pure solution of DNA, the simplest method of quantitation is reading the absorbance at 260 nm where an OD of 1 in a 1 cm path length = 50 mg/ml for double-stranded DNA, 40 mg/ml for single-stranded DNA and RNA and 20 to 33 mg/ml for oligonucleotides. An absorbance ratio of 260 nm and 280 nm gives an estimate of the purity of the solution. Pure DNA and RNA solutions have OD260/OD280 values of 1.8 and 2.0, respectively. This method is not useful for small quantities of DNA or RNA (<1 mg/ml).

(ii) Ethidium bromide fluorescence : The amount of DNA in a solution is proportional to the fluorescence emitted by ethidium bromide in that solution. Dilutions of an unknown DNA in the presence of 2 mg/ml ethidium bromide are compared to dilutions of a known amount of a standard DNA solution spotted on an agarose gel or Saran Wrap or electrophoresed in an agarose gel.

(d) Concentration : Precipitation with ethanol DNA and RNA solutions are concentrated with ethanol as follows: the volume of DNA is measured and the monovalent cation concentration is adjusted. The final concentration should be 2 to 2.5 M for ammonium acetate, 0.3 M for sodium acetate, 0.2 M for sodium chloride, and 0.8 M for lithium chloride. The ion used often depends on the volume of DNA and the subsequent manipulations; for example, sodium acetate inhibits Klenow, ammonium ions inhibit T4 polynucleotide kinase, and chloride ions inhibit RNA-dependent DNA polymerases. The addition of MgC12 to a final concentration of 10 mM assists in the precipitation of small DNA fragments and oligonucleotides. Following addition of the monovalent cations, 2 to 2.5 volumes of ethanol are added, mixed well, and stored on ice or at –20°C for 20 minutes to 1 hour. The DNA is recovered by centrifugation in a microfuge for 10 minutes (room temperature is okay). The supernatant is carefully decanted making certain that the DNA pellet, if visible, is not discarded (often the pellet is not visible until it is dry). To remove salts, the pellet is washed with 0.5 to 1.0 ml of 70% ethanol, spun again, the supernatant is decanted, and the pellet dried. Ammonium acetate is very soluble in ethanol and is effectively removed by a 70% wash. Sodium acetate and sodium chloride are less effectively removed. For fast drying, the pellet can be spun briefly in a Speedvac, although the method is not recommended for many DNA preparations, because DNA that has been over dried is difficult to resuspend and also tends to denature small fragments of DNA. Isopropanol is also used to precipitate DNA but it tends to coprecipitate salts and is harder to evaporate since it is less volatile. However, less isopropanol is required than ethanol to precipitate DNA, and it is sometimes used when volumes must be kept to a minimum, e.g., in large-scale plasmid preps.

(e) **Restriction Enzymes :** Restriction and DNA modifying enzymes are stored at –20°C in a non–frost-free freezer, typically in 50% glycerol. The enzymes are stored in an insulated cooler, which will keep the enzymes at –20°C for some period of time.

GOOD LABORATORY PRACTICES

1. Use a pipette for all the solution transfers. No mouth pipetting.
2. Periodically, clean out refrigerators and deep-freezer.
3. Avoid smoking, eating and drinking in the laboratory.
4. Sterilize all the plastic ware before discarding it.
5. Use gloves and shields. Any radiation is harmful for biological systems. Hard beta from P^{32} is very penetrating.
6. Use glass or plastic goggles while viewing in UV light. Turn UV on for just the minimum required time. UV is damaging to skin and eye.
7. Use gloves while handling ethidium bromide (EtBr). EtBr is highly carcinogenic. Where possible, avoid exposure of nucleic acids to ultraviolet light and also visible light where ethidium bromide is present. This is most important at intermediate steps. Avoid contact with hand and skin.
8. Wash hands thoroughly with soap before leaving the laboratory and before taking any food.
9. Use gloves and sterile tubes, tips, etc. for the success of experiment. Fingers contain nucleases, and dust contains microorganisms, heavy metals etc. all of which degrade nucleic acids.
10. Avoid contact with skin or inhalation of chemicals like Acrylamide (neurotoxin), phenol (corrosive), chloroform, lysozyme, etc.

2 BIOCHEMISTRY

CARBOHYDRATES DETECTION BY MOLISCH TEST

The Molisch test is generally performed for the presence of carbohydrates. Molisch reagent is a solution of á-naphthol in 95% ethanol. This test is useful for identifying any compound that can be dehydrated to furfural or hydroxymethylfurfural in the presence of H_2SO_4. Furfural is derived from the dehydration of pentoses and pentosans, while hydroxymethylfurfural is produced from hexoses and hexosans. Oligosaccharides and polysaccharides are hydrolyzed to yield their repeating monomers by the acid. The α-naphthol reacts with the cyclic aldehydes to form purple condensation products. Though this test will detect compounds other than carbohydrates (i.e., glycoproteins), a negative result indicates the absence of carbohydrates.

Method

Two drops of Molisch reagent is added to 2 ml of the sugar solution and mix thoroughly. Incline the tube, and gently pour 5 ml of concentrated H_2SO_4 down the side of the test tube. A purple color at the interface of the sugar and acid indicates a positive test. Disregard a green color if it appears.

TEST FOR REDUCING SUGARS (BENEDICT'S TEST)

Alkaline solutions of copper are reduced by sugars that have a free aldehyde or ketone group, with the formation of colored cuprous oxide.

Benedict's solution is composed of copper sulfate, sodium carbonate, and sodium citrate (pH 10.5). The citrate will form soluble complex ions with Cu^{++}, preventing the precipitation of $CuCO_3$ in alkaline solutions.

Method

One ml of the solution is added which is to be tested to 5 ml of Benedict's solution, and shake each tube. Place the tube in a boiling water bath and heat for 3 minutes. Remove the tubes from the heat and allow them to cool. Formation of a green, red, or yellow precipitate is a positive test for reducing sugars.

TEST FOR STARCH AND GLYCOGEN (IODINE TEST)

The Lugol's iodine reagent is useful to distinguish starch and glycogen from other polysaccharides. Lugol's iodine yields a blue-black color in the presence of starch. Glycogen reacts with Lugol's reagent to produce a brown-blue color. Other polysaccharides and monosaccharides yield no color change; the test solution remains the characteristic brown-yellow of the reagent. It is thought that starch and glycogen form helical coils. Iodine atoms can then fit into the helices to form a starch-iodine or glycogen-iodine complex. Starch in the form of amylose and amylopectin has less branches than glycogen. This means that the helices of starch are longer than glycogen, therefore binding more iodine atoms. The result is that the color produced by a starch-iodine complex is more intense than that obtained with a glycogen-iodine complex.

Method

Add 2 to 3 drops of Lugol's iodine solution to 5 ml of solution to be tested. Starch produces a blue-black color. A positive test for glycogen is a brown-blue color. A negative test is the brown-yellow color of the test reagent.

IODINE TEST FOR STARCH

The α–4$\rightarrow$ linkages between carbons in the starch produce the helical structure of the polysaccharide chain. The inner diameter of the helix is big enough for elementary iodine to become deposited, thus forming a blue complex (evidence for starch). When starch is mixed with iodine in water, an intensely colored starch/iodine complex is formed. Many of the details of the reaction are still unknown. But it seems that the iodine gets stuck in the coils of β-amylose molecules (β-amylose is a soluble starch). The starch forces the iodine atoms into a linear arrangement in the central groove of the amylose coil. There is some transfer of charge between the starch and the iodine. That changes the way electrons are confined, and so, changes spacing of the energy levels. The iodine/starch complex has energy level spacings that absorb visible light giving the complex its intense blue color.

ESTIMATE THE AMOUNT OF REDUCING SUGARS

To test the presence of reducing sugars in a solution, Benedict's test can be carried out. In Benedict's test, equal volume of Benedict's solution should be added into the unknown solution. The mixture is then boiled. If reducing sugars are present, there should be somc brick-red precipitates. This method can be done quantitatively. The amount of precipitates can indicate the amount of reducing sugars. The more precipitates that form, the more reducing sugars are present in the solution. So by comparing the amount of precipitates formed in the grape juice with the amount of precipitates formed in a standard sugar solution, one can estimate the amount of reducing sugars in a grape. In this experiment, a standard glucose solution of concentration 0.01M was used. It was diluted into different concentrations. The amount of reducing sugars was determined by Benedict's test and the results were plotted in a standard curve.

Procedure

1. Prepare 1 ml of glucose solutions of different concentrations from a stock 0.01 M glucose solution by adding suitable amounts of distilled water. The dilution table is shown:

Molarity of glucose solution (M)	0.000	0.002	0.004	0.006	0.008	0.010
Volume of stock solution added (ml)	0.0	0.2	0.4	0.6	0.8	1.0
Volume of distilled water added (ml)	1.0	0.8	0.6	0.4	0.2	0.0

2. Peel off the epidermis of the grape, put the grape into a beaker and grind the grape tissue with a glass rod to get as much juice as possible, pour the juice into a measuring cylinder and add distilled water into it up to 25 ml.
3. Pipette 1 ml of the prepared diluted juice into a test tube G. Pipette 1 ml of the sugar solutions into test tubes A to F.
4. Place 2 ml of Benedict's solution into each of test tubes A to G. Boil all tubes in water bath for 5 minutes. Allow the precipitates to settle for 10 minutes. Estimate the amount of precipitates in each tube.

DETERMINATION OF BLOOD GLUCOSE BY HAGEDORN-JENSON METHOD

Principle

Blood proteins are precipitated with zinc hydroxide. The reducing sugars in the protein free filtrate reduce potassium ferricyanide on heating. The amount of unreduced ferricyanide is determined iodimetrically.

Reagents

1. **0.45% zinc sulfate :** prepared fresh every week by dilution of 45% stock solution.
2. **0.1N NaOH :** prepared fresh every week by dilution of 2N NaOH. Potassium ferricyanide solution: 1.65 gm crystallized potassium ferricyanide and 10.6 gm anhydrous sodium carbonate were dissolved in 1 liter of water and stored in a dark bottle, protected from light.
3. **Iodide-Sulfate-Chloride Solution :** zinc sulfate 10 gm and NaCl 50 gm were dissolved in 200 ml of water.
4. On the day of use, 5 gm potassium iodide is added to the solution.
5. 3% acetic acid.
6. **0.005N sodium thiosulfate :** prepared fresh daily by diluting 0.5N $Na_2S_2O_3$ was prepared by dissolving 70 gm of salt in 500 ml water. It was better if a solution of slightly higher normality was prepared since sodium thiosulfate decomposes rapidly. The solution should be protected from light and stored in the cold. The normality was checked daily with 0.005N potassium iodate.

7. **0.005N potassium iodate :** This is a stable solution and should be made accurately. 0.3566 gm of the anhydrous salt was weighed accurately and dissolved in 2 liters of H_2O. That was used to check $Na_2S_2O_3$ and potassium ferricyanide solution.
8. **Starch indicator :** 1 gm soluble starch was dissolved in 100 ml saturated NaCl solution.

Procedure

1. One ml 0.1N NaOH and 5 ml 0.45% zinc sulfate were pipetted into a test tube. 0.1 ml blood was taken in a dry pipette and introduced into the gelatinous zinc hydroxide in the test tube, rinsing out the pipette twice with the mixture. The tube was kept in boiling water for 3 minutes and cooled without disturbing the precipitate. The mixture was filtered through a Whatman filter paper of lightly pressed moisture cotton.
2. The tube was washed twice with 3 ml portions of water and filtered into the same containers.
3. Two ml potassium ferricyanide was then added to the filtrate and heated in boiling water for 15 minutes. After cooling, 3 ml iodide-sulfate-chloride solution, followed by 2 ml 3% acetic acid, was added.
4. The liberated iodine was then titrated against 0.005N $Na_2S_2O_3$ using 2 to 3 drops of 1% starch as an indicator toward the end of the titration.
5. A blank was run through the entire procedure simultaneously. The blank should yield a liter value of 1.97 to 2.00 ml.

ESTIMATION OF PROTEIN BY THE BIURET METHOD

Principle

The commonly used method based on the fact that the CO-NH (Peptide) group of proteins forms a purple complex with copper ions in an alkaline medium. Since all proteins contain the peptide bond, the method is fairly specific and there is little interference from other compounds. Some substances like urea and biuret interfere because they possess the CO-NH group. Other interfering materials are reducing sugar like glucose, which interacts with Cu^{+3} ions (cupric) in the reagent.

Reagents Required

1. **Biuret reagent :** Dissolve 1.5 gm of $CuSO_4$ and 4.5 gm of Na-K-tartrate in 250 ml 0.2 N NaOH solution. Add 2.5 gm of KI and make up the volume to 500 ml with 0.2 N NaOH.
2. 0.2 N NaOH.
3. **Protein standard solution :** Dissolve 500 mg of egg albumin in 50 ml of H_2O. Make up the volume to 100 ml to get the final concentration of 5 mg/ml.

Procedure

Pipette out standard protein solution into a series of tubes 0.0, 0.2, 1 ml and make up the total volume to 4 ml by adding water. The blank tube will have only 4 ml of water. Add 6 ml of biuret reagent to each tube and mix well. Keep the tubes at 37°C for 10 minutes during which a purple color will develop. The optical density of each tube is measured at 52.0 nm (green filter) using the blank reagent. Draw the graph to the known concentrate of a protein in an unknown solution.

ESTIMATION OF PROTEIN BY THE LOWRY PROTEIN ASSAY

Materials

1. 0.15% (w/v) sodium deoxycholate
2. 72% (w/v) Trichloroacetic acid (TCA)
3. Copper tartrate/carbonate (CTC)
4. 20% (v/v) Folin-Ciocalteu reagent
5. Bovine serum albumin (BSA)
6. Spectrophotometer and tubes
7. Micropipettes

Procedure

1. Prepare standard dilutions of BSA of 25, 50, 75, and 100 mg/ml. Prepare appropriate serial dilutions of the sample to be measured.

2. Place 1.0 ml of each of the above into separate tubes. Add 100 ml of sodium deoxycholate to each tube.
3. Wait 10 minutes and add 100 ml of TCA to each tube.
4. Centrifuge each tube for 15 minutes at 3000 rpm and discard the supernatant.
5. Add 1.0 ml of water to each tube to dissolve the pellet. Add 1.0 ml of water to a new tube to be used as a blank.
6. Add 1.0 ml of CTC to each tube (including the blank), vortex and allow to set for 10 minutes.
7. Add 500 ml Folin-Ciocalteu to each tube, vortex and allow to set for 30 minutes.
8. Turn on and zero a spectrophotometer to a wavelength of 750 nm. Use the blank from step 7 to adjust for 100% T.
9. Read each of the standards and samples at 750 nm.
10. Plot the absorbance of the standards vs their concentration. Compute the extinction coefficient and calculate the concentrations of the unknown samples.

ESTIMATION OF DNA BY THE DIPHENYLAMINE METHOD

Principle

When a sample of DNA is treated with diphenylamine under acidic conditions, a blue compound is formed with the sharp observation maximum at 595 nm. This reaction is produced by 2 deoxypentose in general and is not specific for DNA. In acid solution the straight from the deoxypentose is converted to the highly reactive α-hydroxy lerulin aldehyde, which reacts with diphenylamine to produce a blue complex. In DNA, only the deoxyribose of purine nucleotide reacts so that the value obtained represents one half of the total deoxyribose produced.

Materials

1. Take 10 mg sample of DNA.
2. Buffer saline 0.15 ml/L NaCl, 0.15 ml/L sodium citrate made up to 500 ml.

3. Diphenyl amine reagent: dissolved in 10 gm of pure diphenyl amine in 1 liter of glacial acetic acid and add 25 ml of concentrated H_2SO_4. This solution must be prepared fresh.
4. Boiling water bath.

Procedure

Dissolve 10 gm of nucleic acid in 50 ml of buffer saline. Remove 2 ml and add 4 ml of diphenylamine. Remove 2 ml and add 4 ml of DPA reagent. Heat on a boiling water bath for 10 minutes. Cool and read the extension at 595 nm. Read the test and standard against nucleic acid and the commercial sample for DNA.

ESTIMATION OF RNA BY THE ORCINOL METHOD

Principle

This is a general reaction for pentose and depends on the formation of furfural. When pentose is heated with concentration HCl, orcinol reacts in presence of furfural in presence of ferric chloride as a catalyst purine to produce green color only the purine nucleotide.

Materials

1. Take 10 mg RNA sample.
2. **RNA solution (0.2 mg/ml) :** Dissolve 10 mg of RNA in 50 ml of buffer saline and use the sample for standard preparation.
3. **Orcinol reagent :** Dissolve 10 gm of ferric chloride $(FeCl_3)_6$ H_2O in 1 liter of 1 gm of ferric chloride $(FeCl_3)_6$ H_2O in 1 liter of concentrated HCl and add 35 ml of 6% w/v orcinol in alcohol.
4. Buffered saline.
5. Boiling water bath.

Procedure

Pipette out standard RNA solution in arrange of 0 to 2 ml into a series of test tube and make up the volume of each tube 2 ml with distilled water. Add 3 ml of orcinol reagent to each tube for standard solution. For test solution: Take 2 ml of nucleic acid sodium (isolated from tissue source.) Add 3 ml of orcinol reagent to each tube, and heat the tube on boiling water bath for 20 minutes. Cool and take the optical density at 665 nm against the orcinol blank.

HIGH PRESSURE LIQUID CHROMATOGRAPHY (HPLC)

Introduction

HPLC is a form of column chromatography used frequently in biochemistry and analytical chemistry to separate, identify and quantify compound .HPLC utilize a column that holds chromatography packing materials (stationary phase). A pump that moves the mobile phase through the column, and a detector that show the retention times of the molecules. Retention time varies depending on the interaction between the stationary phases. The molecules being analyzed and the solvent used.

Component of HPLC

The solvent reservoir

It is generally made of glass. Before use it is necessary to degas the solvent to remove dissolved gases (particularly oxygen), which may interact either with mobile or stationary phase.

Pumping system

Different types of pumps are used for solvent delivery system e.g. constant volume type and constant type.

Flow controller

These consists of flow through pressure transducer which measures the flow rate by measuring the pressure drop across restrictor of fixed value, placed at the pump outlet. The flow rate signal is back to a control unit which compares the actual and pre set flow rate.

Sample injectors

The sample is introduced normally by syringe through the septum of an injection port into the system by using an auto injector.

Detector system

An ideal detector should have a high sensitivity (be able to detect less than 1 ppm concentration,) non destructive. Most commonly used detector is UV-visible detector (170 to 700 nm) and photodiode array detector.

Sample filtration

Before injection the sample is passed through the sample filter (0.5 μm nylon membranes) to assume complete particulate removal and maximum column life. Filtration prevents undesirable contaminants from clogging valuables columns

Solvent/ eluent clarification

The eluent was filtered using solvent filtration kit. This removes the particulates very quickly and easily. Solution is deaerated for detector stability as air bubbles may cause spikes or noise in detectors. Mostly nylon membranes (0.25 or 00.5 μm) are used they can filter both the hydrophobic and hydrophilic solvents.

Materials

1. Acetonitrile
2. Methanol
3. Acetic acid
4. Sterile distill water

Methodology

(To use Empower software startup guide line)

1. Click on Empower window.
2. After window comes click ok

3. Click in run window & wait for open the window (if you want create new method than go to edit tab & below window will open. click on W2996. click 2D data collection & feed desired wavelength as 272).
4. Go to PCM /15 windows as below & click on flow & feed the desired flow rate, high, low pressure limit & feed the desired % of A.B.
5. After all parameter go to file save as & feeds the instrument method name. After that go to exit.
6. In main window feed the sample name, choose standard & unknown as your requirement & select method set.
7. Feed the vial no. injected volume & desired run time. After that click on inject & wait till your run complete.
8. After that completed run click Empower window & window will appear as below click on browse project, select desire project & click ok. Below window will appear, click in injection tab and select the desired sample which is you want to process.
9. Click right to review & go to review window will appear as below.
10. In method wizard window you have to click ok & another window will appear as below.
11. As mentioned above as chromatogram zoom the interest peak area & click next. In this window zoom the less noise area & click next. If you do not want small 2 peaks than minimum area & for checking click test & click next.
12. Click next and yes. Give the interested peaks name as below & click next-2 Give the processing method name as below & click finish.
13. After save the method click on integrate Tab, calibrate & save all tab.
14. After that you have to click in results. Click on demo window tab & click in result tab, after that select the desired sample & right click, which is saved after processing. Go to preview/Publisher & as below window with appear.
15. Click ok. If you want to print report than you have to click printer tab menu as shown. Creating method set go to review window method set below window will appear here you have to select instrument method processing method go to channel select desired wavelength processing method as shown below window go to file save as give method set name exit.

16. For calculating the unknown value with respect to standard run four five standard & select all standard right click go to after sample. Below window will appear & go o edit amount.
17. As below window will appear go to again edit copy component from process method. Below window will appear & this will take peak name from processing method.
18. Feed the desire value of concentration & click ok. Below window will appear save exit from window.
19. Select all standard right click process select desired method set click clear. calibration ok go to result tab preview/ publisher select LC calibration report format click ok you will gate proper report format.

GAS CHROMATOGRAPHY (GC)

Introduction

A common type of chromatography which is used in analytic chemistry for separating and analyzing compounds that can be vaporized without decomposition. Typical uses of GC include testing the purity of a particular substance, or separating the different components of a mixture (the relative amounts of such components can also be determined). In some situations, GC may help in identifying a compound. In preparative chromatography, GC can be used to prepare pure compounds from a mixture.

Principle

In this type of chromatography, the moving phase (or "mobile phase") is a carrier gas, usually an inert gas such as helium or an unreactive gas such as nitrogen. The stationary phase is a microscopic layer of liquid or polymer on an inert solid support, inside a piece of glass or metal tubing called a column (an homage to the fractionating column used in distillation). The instrument used to perform gas chromatography is called a gas chromatograph (or "aerograph", "gas separator").

The gaseous compounds being analyzed interact with the walls of the column, which is coated with different stationary phases. This causes each compound to elute at a different time, known as the retention time of the compound. The comparison of retention times is what gives GC its analytical usefulness.

Gas chromatography is in principle similar to column chromatography (as well as other forms of chromatography, such as HPLC, TLC), but has several notable differences. Firstly, the process of separating the compounds in a mixture is carried out between a liquid stationary phase and a gas moving phase, whereas in column chromatography the stationary phase is a solid and the moving phase is a liquid. (Hence the full name of the procedure is "Gas-liquid chromatography", referring to the mobile and stationary phases, respectively.) Secondly, the column through which the gas phase passes is located in an oven where the temperature of the gas can be controlled, whereas column chromatography (typically) has no such temperature control. Thirdly, the concentration of a compound in the gas phase is solely a function of the vapor pressure of the gas.

Gas chromatography is also similar to fractional distillation, since both processes separate the components of a mixture primarily based on boiling point (or vapor pressure) differences. However, fractional distillation is typically used to separate components of a mixture on a large scale, whereas GC can be used on a much smaller scale (i.e. microscale).

Component of gas Chromatography

Injector

The sample is introduced normally by syringe through the septum of an injection port into the system by using an auto injector.

Coloumn

Commonly open tubular capillary used. Column "frame" constructed of fused silica tubing and Polyamide coating on the outside gives it strength. In column Liquid stationary phases coated or bonded to the inside of the tubing. 0.1 to 0.53 mm + ID, 5-100 meters in length, stationary phases usually 0.10 to 1.5 mm in thickness. Mounted on a wire cage to make them easier to handle mostly columns are 5 to 150 meters long.

Oven

Oven is for maintaining the temperature of column.

Detector system

An ideal detector should have a high sensitivity (be able to detect less than 1 ppm concentration,) non destructive. Most commonly used detector is flame ionization detector. Alkali(FID), Flame Ionization Detector (AFID), Nitrogen Phosphorus Detector (NPD), Electron Capture Detector (ECD), Flame Photometric Detector (FPD), Mass Detector (MS).

Material

1. Sample
2. Gases: nitrogen, hydrogen, air

Methodology

1. Start total chrom by double clicking the TC Naviem. Enter username password when pumped.
2. From the navigator screen, close system configuration from the admin menu.
3. When you open the system configuration editor, total chrom display the current system configuration.
4. Select the ports tab.
5. Select add acquisitions port from the ports menu to open the add acquisitions port b dialog box.
6. Select serial from the Device.
7. Create new method with the help of software and develop method as per required.
8. Wait till the similar protocol sets on the Instrument as well.
9. After the instrument has set the protocol Run the Sample and wait for the injection to run.
10. The relevant peak will be observer on the screen which could be viewed by clicking the Report format edit.
11. This Raw data then need to be processed by calibrating, naming and monitoring the baseline components.
12. Then the processes data can be reviewed by taking its Print.

3 MICROBIOLOGY

INTRODUCTION

The microscope has been a well known and a valuable tool in the development of scientific theory and study of cells, microbes, etc., there are various types of microscopes available depending upon their use and functionality. A compound microscope is composed of 2 elements, a primary magnifying lens and a secondary lens system. Light passes through an object and is then focused by the primary and secondary lens. If the beam of light is replaced by an electron beam, the microscope becomes a transmission electron microscope. If light is bounced off of the object instead of passing through, the light microscope becomes a dissecting scope. If electrons are bounced off of the object in a scanned pattern, the instrument becomes a scanning electron microscope.

The function of any microscope is to enhance resolution. The microscope is used to create an enlarged view of an object so that we can observe details not otherwise possible with the human eye. Because of the enlargement, resolution is often confused with magnification, which refers to the size of an image. In general, the greater the magnification, the greater the resolution, but this is not always true. There are several practical limitations of lens design that can result in increased magnification without increased resolution.

If an image of a cell is magnified from 10X to 45X, the image gets larger, but not necessarily any clearer. The image on the left is magnified with no increase in resolution. The image on the right is magnified the same, but with increasing resolution. Note that by the time the image is magnified 10X (from 10X to 100X), the image on the left is completely unusable. The image on the right, however, presents more detailed information. Without resolution, no matter how much the image is magnified, the amount of observable detail is fixed, and regardless of how much you increase the size of the image, no more detail can be seen. At this point, you will have reached the limit of resolution or the resolving power of the lens. This property of the lens is fixed by the design and construction of the lens. To change the resolution, a different lens is often the only answer.

The reason for a dichotomy between magnification and resolution is the ability of the human eye to see 2 objects. It is necessary that 2 objects be about 0.1-mm apart when held 10" from the face in order for us to detect them as 2 objects. If they are closer than 0.1 mm, we will perceive them as a single object. If 2 objects are 0.01-mm apart, we cannot detect them unless we magnify an image of them by 10X. What has happened is that we have effectively altered our resolution ability from 0.1 mm to 0.01 mm through the use of a magnifying lens. Our limit of resolution has changed from 0.1 mm to 0.01 mm, or inversely, our resolving power (resolution) has increased by a factor of 10.

METHOD TO USE MICROSCOPE

1. **Moving and transporting the microscope :** Grasp the arm of the microscope with one hand and support the base of the microscope with the other. Handle it gently.
2. Before you plug in the microscope, turn the voltage control dial on the right side of the base of the microscope to 1. Now plug in the microscope and use the on/off switch in the front on the base to turn it on. Make sure the entire cord is on the bench top and not hanging down where it could be caught on a leg. Adjust the voltage control dial to 10.
3. **Adjusting the eyepieces :** These microscopes are binocular, i.e., they have 2 ocular lenses (eyepieces). To adjust them, first find the proper distance between your eyes and the eyepieces by closing one eye and slowly moving your head toward that eyepiece until you see

the complete field of view about 1 inch away. Keep your head steady and both eyes in the same plane. Now open the other eye and gradually increase the distance between the eyepieces until it matches thc distance between your eyes. At the correct distance, you will see one circular field of view with both eyes.

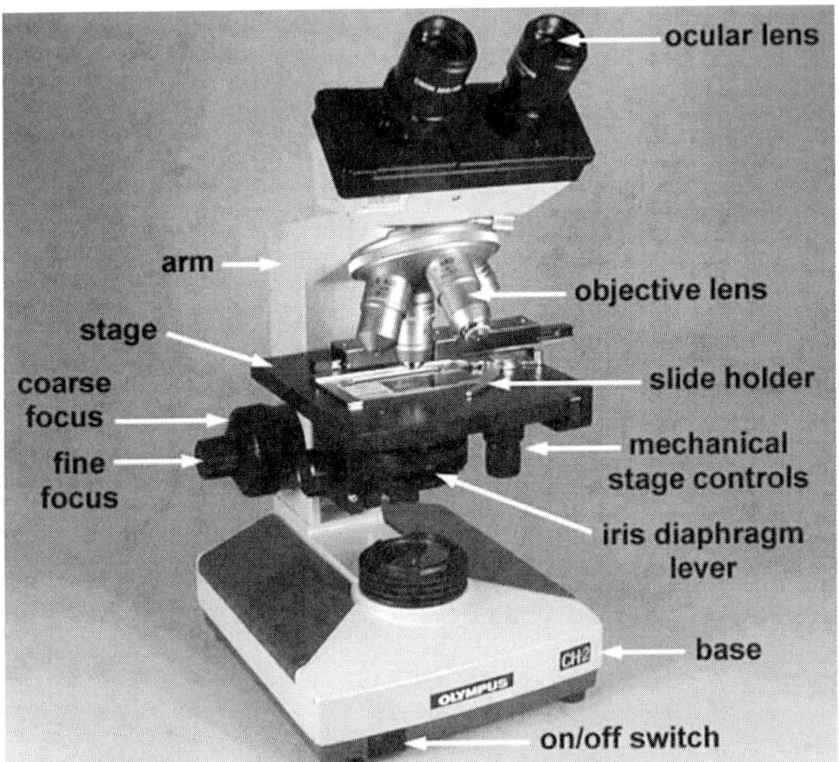

Fig. 1. Olympus microscope.

4. **Positioning the slide :** Place the slide specimen-side-up on the stage so that the specimen lies over the opening for the light in the middle of the stage. Secure the slide between, not under, the slide holder arms of the mechanical stage. The slide can now be moved from place to place using the 2 control knobs located under the stage on the right of the microscope.

5. **Adjusting the illumination :**
   (a) Adjust the voltage by turning the voltage control dial located in the rear right hand side of the microscope base. For oil immersion microscopy (1000X), set the light on 9 or 10. At lower magnifications, less light will be needed.
   (b) Adjust the amount of light coming through the condenser using the iris diaphragm lever located under the stage in the front of the microscope. Light adjustment using the iris diaphragm lever is critical to obtaining proper contrast. For oil immersion microscopy (1000X), the iris diaphragm lever should be set almost all the way open (to your left for maximum light). For low powers such as 100X, the iris diaphragm lever should be set mostly closed (to your right for minimum light).
   (c) The condenser height control (the single knob under the stage on the left hand side of the microscope) should be set so the condenser is all the way up.
6. **Obtaining different magnifications :** The final magnification is a product of the 2 lenses being used. The eyepiece or ocular lens magnifies 10X. The objective lenses are mounted on a turret near the stage. The small yellow-striped lens magnifies 10X, the blue-striped lens magnifies 40X, and the white-striped oil immersion lens magnifies 100X. Final magnifications are as follows:

Ocular lens	×	Objective lens	=	Total magnification
10X	×	10X (yellow)	=	100X
10X	×	40X (blue)	=	400X
10X	×	100X (white)	=	1000X

7. **Focusing from lower power to higher power :**
   (a) Rotate the yellow-striped 10X objective until it locks into place (total magnification of 100X).
   (b) Turn the coarse focus control (larger knob) all the way away from you until it stops.
   (c) Look through the eyepieces and turn the coarse focus control (larger knob) toward you slowly until the specimen comes into focus.

(d) Get the specimen into sharp focus using the fine focus control (smaller knob) and adjust the light for optimum contrast using the iris diaphragm lever.

(e) If higher magnification is desired, simply rotate the blue-striped 40X objective into place (total magnification of 400X) and the specimen should still be in focus. (Minor adjustments in fine focus and light contrast may be needed.)

(f) For maximum magnification (1000X or oil immersion), rotate the blue striped 40X objective slightly out of position and place a drop of immersion oil on the slide. Now rotate the white-striped 100X oil immersion objective into place. Again, the specimen should remain in focus, although minor adjustments in fine focus and light contrast may be needed.

8. **Cleaning the microscope :** Clean the exterior lenses of the eyepiece and objective before and after each lab using lens paper only. (Paper towels may scratch the lens.) Remove any immersion oil from the oil immersion lens before putting the microscope away.

9. Reason for using immersion oil. Normally, when light waves travel from one medium into another, they bend. Therefore, as the light travels from the glass slide to the air, the light waves bend and are scattered, similar to the "bent pencil" effect when a pencil is placed in a glass of water. The microscope magnifies this distortion effect. Also, if high magnification is to be used, more light is needed. Immersion oil has the same refractive index as glass and, therefore, provides an optically homogeneous path between the slide and the lens of the objective. Light waves thus travel from the glass slide, into glass-like oil, into the glass lens without being scattered or distorting the image. In other words, the immersion oil "traps" the light and prevents the distortion effect that is seen as a result of the bending of the light waves.

BRIGHT FIELD, DARK FIELD, PHASE CONTRAST

All microscopes actually allow visualization of objects through minute shifts in the wavelength phase as the light passes through the object. Further image forming can be had through the use of color, or through a complete negative image of the object. If the normal phase shift is increased (usually by 1/4 wavelength), then the microscope becomes a phase contrast microscope. Phase contrast microscopes can be designed to have medium-phase or dark-phase renditions, by altering the degree of additional shift to the wavelength from ¼ to ½ wavelengths, respectively. If the beam of light is shifted in phase by a variable amount, the system becomes a differential interference contrast microscope. If the light image is reversed, then the microscope becomes a dark-field microscope. All standard bright-field microscopes can be readily converted to dark-field by inserting a round opaque disk beneath the condenser. Dark-field microscopy was first utilized to examine trans-filterable infectious agents, later to be termed viruses, and to determine that they were particulate in nature. Small objects, even those below the limits of resolution, can be detected easily with dark-field, as the object appears to emit light on a dark-field. (Look at the sky for a comparison. It is fairly easy to see stars in a dark sky, but impossible during the day. The same is true for dark-field versus bright-field microscopy). Finally, if the normal light microscope is functionally turned upside down, the microscope becomes an inverted microscope. This is particularly useful in tissue culture, since it allows observation of cells through the bottom of a culture vessel, without opening the container, and without the air interface normally present between the objective and the surface of the culture. By adding phase contrast optics to the inverted microscope, it is possible to monitor tissue cultures directly, without the aid of stains or other enhancements.

THE ELECTRON MICROSCOPE

A. **The transmission electron microscope (TEM) :** has resolving power (3 to 10 Å). The scanning electron microscope (SEM) is becoming increasingly popular with cell biologists because of its remarkable ability for quantifiable mapping of surface detail, along with improved resolution (30 to 100 Å) and its ability to show 3D structure. The transmission electron microscope is identical in concept to the modern binocular light microscope. It is composed of a light source (in this case an electron source), a sub stage condenser to focus the electrons on the specimen, and an objective and ocular lens system. In the electron microscope, the ocular lens is replaced with a projection lens, since it projects an image onto a fluorescent screen or a photographic plate. Since the electrons do not pass through glass, they are focused by electromagnetic fields. Instead of rotating a nosepiece with different fixed lenses, the EM merely changes the current and voltage applied to the electromagnetic lenses. The size of an electron microscope is dependent upon 2 factors. The first is the need for a good vacuum through which the electrons must pass (it takes less than 1 cm of air to completely stop an electron beam). Peripheral pumps and elaborate valve controls are needed to create the vacuum. A substantial electrical potential (voltage) is also needed to accelerate the electrons out of the source. The source is usually a tungsten filament, very much like a light bulb, but with 40 to 150 kilovolts of accelerating voltage applied to an anode to accelerate the electrons down the microscope column. Modern electronics have produced transformers that are reasonably small but capable of generating 60,000 volts. Another characteristic of electron microscopes is that they are usually designed upside down, similar to an inverted light microscope. The electron source is on top, and the electrons travel down the tube, opposite light rays traveling up a microscope tube. This is merely a design feature that allows the operator and technicians ease of access to its various components. The newer electron microscope is beginning to look like a desk with a TV monitor on it. Until recently, the major advantage of an electron microscope has also been its major disadvantage. In theory, the transmission electron microscope should be capable of producing a resolution of several angstroms. This would provide excellent molecular resolution of cell organelles. However, as the resolution increases, the field of view decreases and it becomes increasingly difficult to view the molecular detail within the cell. Electron microscopes designed to yield high resolution have to be compromised to view larger objects. Cell structures fall within the size range that was most problematic for viewing. For example, if we wished to resolve the architecture of an entire eukaryotic chromosome, not just the chromosome, but the cell itself was too large to be seen effectively in an electron microscope. Zooming in on paired chromosomes was impossible. Modern electron microscope design allows for this zooming, and for the observation of whole tissues while retaining macromolecular resolution.

B. The Scanning Electron Microscope : The scanning electron microscope works by bouncing electrons off of the surface and forming an image from the reflected electrons. Actually, the electrons reaching the specimen are normally not used (although they can form a transmitted image, similar to standard TEM), but they incite a second group of electrons to be given off from the very surface of the object. Thus, if a beam of primary electrons is scanned across an object in a raster pattern (similar to a television scan), the object will give off secondary electrons in the same scanned pattern. These electrons are gathered by a positively charged detector, which is scanned in synchrony with the emission beam scan. Thus, the name scanning electron microscope, with the image formed by the collection of secondary electrons. It is possible to focus the primary electrons in exactly the same manner as a TEM. Since the primary electrons can be focused independently of the secondary electrons, 2 images can be produced simultaneously. Thus, an image of a sectioned material can be superimposed on an image of its surface. The instrument then becomes a STEM, or scanning-transmission electron microscope. It has the same capabilities of a TEM, with the added benefits of an SEM. SEM allows a good deal of analytical data to be collected, in addition to the formed image. As the primary electrons bombard the surface of an object, they interact with the atoms of the surface to yield even more particles and radiations besides secondary electrons. Among these radiations are Auger electrons and characteristic x-rays. The x-rays have unique, discrete energy values, characteristic of the atomic structure of the atom from which they emanated. If one collects these x-rays and analyzes their inherent energy, the process becomes energy-dispersive x-ray analysis. Combining the scan information from secondary and Auger electrons, together with the qualitative and quantitative x-ray information, allows the complete molecular mapping of an object's surface. Finally, the scanning microscope has one further advantage that is useful in cell structure analysis. As the electron beam scans the surface of an object, it can be designed to etch the surface. That is, it can be made to blow apart the outermost atomic layer. As with the emission of characteristic x-rays, the particles can be collected and analyzed with each pass of the electron beam. Thus, the outer layer can be analyzed on the first scan, and subsequently lower layers analyzed, with each additional scan. Electrons are relatively small, and the etching can be enhanced by bombarding the surface with ions rather than electrons (the equivalent of bombarding with bowling balls rather than BBs). The resultant secondary emissions-ion scanning data can finally be analyzed and the 3-dimensional bitmapped atomic image of an object can be reconstructed.

MICROSCOPY

Requirement : Various types of microscopes.

Theory : Magnification of microscope depends on the following factors

1. Optical tube length of the objective
2. The focal length of the objective and
3. Magnification of the eye piece.

$$\text{The primary magnification (by objective lens)} = \frac{\text{Optical tube length}}{\text{Focal length of objective}}$$

Microscope Resolution

The most important part of the microscope is the objective, which must produce a clear image, not just a magnified one. Thus resolution is extremely important. Resolution is the ability of a lens to separate or distinguish between small objects that are close together. Much of the optical theory underlying microscope design was developed by the German physicist Ernst Abbe in the 1870s. The minimum distance (d) between two objects that reveals them as separate entities is given by the Abbe equation, in which lambda (λ) is the wavelength of light used to illuminate the specimen and $n \sin \theta$ is the numerical aperture (NA).

$$d = \frac{0.5(\lambda)}{n \sin \theta}$$

The resolution of a microscope depends upon the numerical aperture of its condenser as well as that of the objective. This is evident from the equation describing the resolution of the complete microscope.

$$\text{D microscope} = \frac{\lambda}{(\text{NA objective} - \text{NA condenser})}$$

Most microscopes have a condenser with a numerical aperture between 1.2 and 1.4. However, the condenser numerical aperture will not be much above about 0.9 unless the top of the condenser is oiled to the bottom of the slide. During routine microscope operation, the condenser usually is not oiled and this limits the overall resolution, even with an oil immersion objective. The limits set on the resolution of a light microscope can be calculated using the Abbe equation.

At best, a bright-field microscope can distinguish between two dots around 0.2 m apart (the same size as a very small bacterium). As d becomes smaller, the resolution increases, and finer detail can be discerned in a specimen. The preceding equation indicates that a major factor in resolution is the wavelength of light used. The wavelength must be shorter than the distance between two objects or they will not be seen clearly. Thus the greatest resolution is obtained with light of the shortest wavelength, light at the blue end of the visible spectrum (in the range of 450 to 500 nm).

Similarly, the various microscope i.e., simple microscope, compound microscope, bright field microscope, fluorescence microscope, electron microscope, TEM & SEM are studied.

Observations and result : As observed in the experiment

TO IDENTIFY AND STAIN DIFFERENT TYPES OF CELLS

Requirements : Chemicals & reagents, glassware's etc.

Method : Acid-Alcohol (for Ziehl-Neelsen stain) 3 ml concentrated hydrochloric acid in 100 ml of 95% ethyl alcohol.

Acid-Alcohol (for fluorochrome staining)

HCl	2.5 ml
Ethyl alcohol, 70%	500.0 ml
NaCl	2.5 gm

Alcohol, 70% (from 95%)

Alcohol, 95%	368.0 ml
Distilled water	132.0 ml

Auramine-Rhodamine Stain (for Mycobacteria)

Auramine 0	3.0 gm
Rhodamine B	1.5 gm
Glycerol	150.0 ml
Phenol crystals (liquefied at 50°C)	20.0 ml
Distilled water	100.0 ml

Clarify by filtration with glass wool or Whatman filter paper. Do not heat. Store at room temperature.

Barritt's Reagent (Voges-Proskauer test)

Solution A : 6 gm alpha-naphthol in 100 ml 95% ethyl alcohol.

Solution B : 16 gm potassium hydroxide in 100 ml water. Note that no creatine is used in these reagents as is used in O'Meara's reagent for the V-P test.

Carbolfuchsin Stain (Ziehl's)

Solution A : Dissolve 0.3 gm of basic fuchsin (90% dye content) in 10 ml 95% ethyl alcohol.

Solution B : Dissolve 5 gm of phenol in 95 ml of water. Mix solutions A and B.

Crystal Violet Stain (Hucker modification)

Solution A : Dissolve 2.0 gm of crystal violet (85% dye content) in 20 ml of 95% ethyl alcohol.

Solution B : Dissolve 0.8 gm ammonium oxalate in 80.0 ml distilled water. Mix solutions A and B.

Nitrate test by Diphenylamine Reagent

1. Add 0.7 gm diphenylamine in a mixture of 60 ml of concentrated sulfuric acid and 28.8 ml of distilled water.
2. Cool and add slowly 11.3 ml of concentrated hydrochloric acid.

3. After the solution has stood for 12 hours some of the base separates, showing that the reagent is saturated.

Ferric Chloride Reagent

$FeCl_3 \cdot 6H_2O$	12 gm
2% Aqueous HCl	100 ml

1. Make up the 2% aqueous HCl by adding 5.4 ml of concentrated HCl (37%) to 94.6 ml H_2O.
2. Inoculate with two or three colonies of β-hemolytic streptococci incubate at 35° C for 20 or more hours.
3. Centrifuge the medium to pack the cells, and pipette 0.8 ml of the clear supernatant into a Kahn tube.
4. Add 0.2 ml of the ferric chloride reagent to the Kahn tube and mix well. If a heavy precipitate remains longer than 10 minutes, the test is positive.

Gram's Iodine (Lugol's) : Dissolve 2.0 gm of potassium iodide in 300 ml of distilled water and then add 1.0 gm iodine crystals.

Iodine, 5% Aqueous Solution : Dissolve 4 gm of potassium iodide in 300 ml of distilled water and then add 2.0 gm iodine crystals.

Kovacs' Reagent (indole test)

n-amyl alcohol	75.0 ml
Hydrochloric acid (conc.)	25.0 ml
Dimethylamine-benzaldehyde	5.0 gm

Lactophenol Cotton Blue Stain

Phenol crystals	20 gm
Lactic acid	20 ml
Glycerol	40 ml
Cotton blue	0.05 gm

Dissolve the phenol crystals in the other ingredients by heating the mixture gently under a hot water tap.

Malachite Green Solution (spore stain) : Dissolve 5.0 gm malachite green oxalate in 100 ml distilled water.

Observations & results : As observed in the experiment

TO MEASURE VARIOUS CELL ORGANELLES

Requirements : sample of any microorganisms, microscope, ocular micrometer etc.

Method : For determining the absolute size of cells it becomes necessary to first measure the length in micrometers (μm) between two lines of the ocular micrometer. For this purpose a stage micrometer with a scale measured in micrometers becomes necessary for μm calibration of the ocular micrometer. The stage micrometer scale is such that the distance between two lines is 0.01 mm (equivalent to 10 μm). By superimposing the stage micrometer scale over the ocular micrometer scale, one can determine absolute values in microns between two lines on the ocular micrometer scale. The absolute value obtained is also dependent on the objective used.

For example, with the low power objective, seven divisions on the ocular micrometer = one division on the stage micrometer (Figure 2). Thus with the low power objective, one division on the ocular micrometer = .01 mm/7 = 1.40 micrometers (μm).

Procedure for Insertion, Calibration, and Use of the Ocular Micrometer

1. Place a clean stage micrometer in the mechanical slide holder of the microscope stage.
2. Using the low power objective, center and focus the stage micrometer.
3. Unscrew the top lens of the ocular to be used, and then carefully place the ocular micrometer with the engraved side down on the diaphragm inside the eyepiece tube (see figure 1), followed by replacing the top lens of the ocular.

   Note : With some microscopes the ocular micrometer is inserted in a retaining ring located at the base of the ocular.
4. To calibrate the ocular micrometer, rotate the ocular until the lines of the ocular micrometer are superimposed over the lines of the stage micrometer.
5. Next move the stage micrometer until the lines of the ocular and stage micrometer coincide at one end.

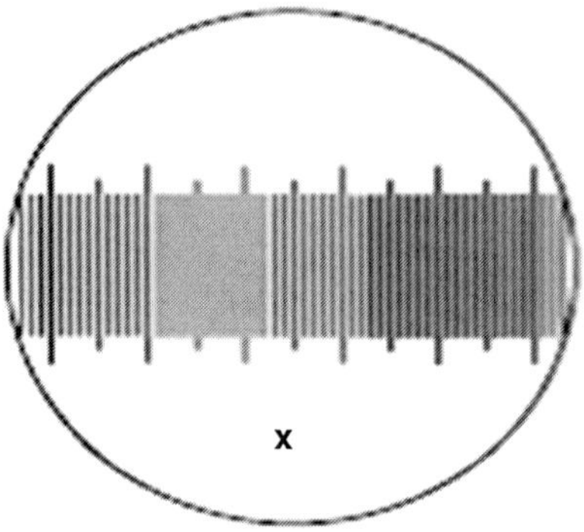

(a) Ocular micrometer The dismeter (wiclth) of the graduations in microns must be determined for each objective.

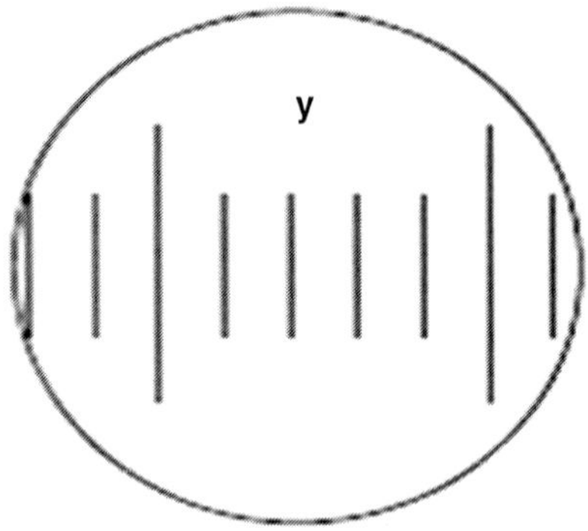

(b) Stage micrometer The graduations are .01mm (10 μm) wide.

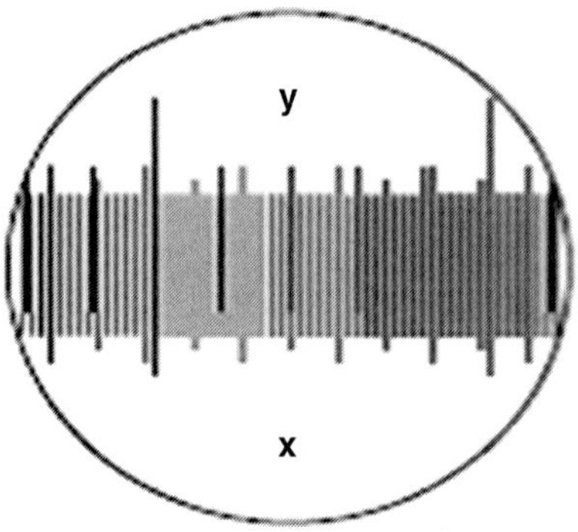

(c) Superimposing of ocular micrometer scale (x) over the stage micrometer scale (y) Note that seven divisions of the ocular micrometer equal one division of the stage micrometer scale (.01mm).

One division of x = $\frac{.01}{7}$

= .0014mm = 1.4 μm

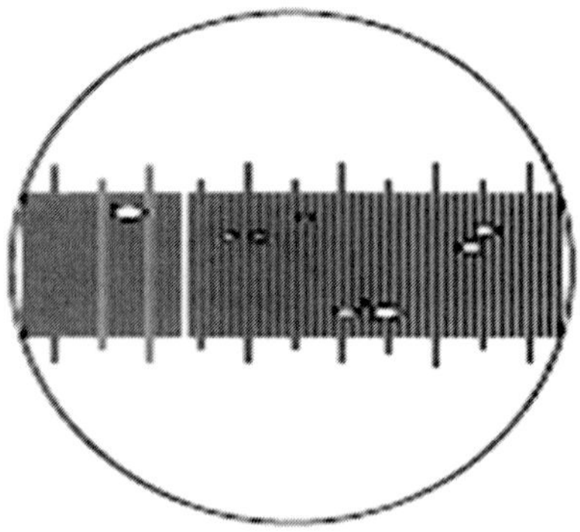

(d) Based on the Figure 2(c) calculations, what is the average length in microns of the rod shaped bacteria?

Fig. 2. Calibration of the ocular micrometer.

6. Now find a line on the ocular micrometer that coincides precisely with a line on the stage micrometer.
7. Determine the number of ocular micrometer divisions and stage micrometer divisions where the two lines coincide. A case in point is shown in figure.

   Observations & results : As observed in the experiment

DETECTION OF BACTERIAL MOTILITY

Requirement: Glassware & bacterial sample.

Introduction: Many bacteria are capable of motility. Most motile bacteria propel themselves by special organelles termed flagella.

The following flagellar arrangements may be found in bacteria:

1. *Monotrichous,* a single flagellum at 1 pole.
2. *Amphitrichous,* a single flagella at both poles.
3. *Lophotrichous,* 2 or more flagella at 1 or both poles of the cell.
4. *Peritrichous,* completely surrounded by flagella.

Organisms: Trypticase soy broth cultures of *Pseudomonas aeruginosa* and *Staphylococcus aureus*.

Motility test medium: A semi-solid motility test medium may also be used to detect motility. The agar concentration (0.3%) is sufficient to form a soft gel without hindering motility. A tetrazolium salt (TTC) is incorporated into the medium. Bacterial metabolism reduces the TTC producing formazan, which is red. The more bacteria present at any location, the darker red the growth appears.

Procedure:

1. Observe the phase-contrast video microscopy demonstration of motile *Pseudomonas aeruginosa*.
2. Observe the dark-field microscopy demonstration of motile *Pseudomonas aeruginosa*.
3. Take 2 tubes of motility test medium per pair. Stab one with *Pseudomonas aeruginosa* and the other with *Staphylococcus aureus*. Incubate at 37°C until the next lab period.

4. Observe the flagella stain demonstrations of *Pseudomonas aeruginosa* (monotrichous), *Proteus vulgaris* (peritrichous), and *Spirillum undula* (lophotrichous), as well as the dark-field photomicrograph of the spirochete.

Observations and Results:

The flagellar movement has to look carefully to observe the true flagellar arrangement.

1. Observe the phase contrast and dark field microscopy demonstrations of bacterial motility.
2. Observe the 2 tubes of motility test medium.
3. Make drawings of the flagella stain demonstrations.

TO DETECT MITOSIS WITH THE HELP OF MICROSCOPE

Mitosis

When a eukaryotic microorganism reproduces, its genetic material must be duplicated and then separated so that each new nucleus possesses a complete set of chromosomes. This process of nuclear division and chromosome distribution in eukaryotic cells is called mitosis. Mitosis actually occupies only a small portion of a microorganism's life as can be seen by examining the cell cycle. The cell cycle is the total sequence of events in the growth division cycle between the end of one division and the end of the next. Cell growth takes place in the interphase, that portion of the cycle between periods of mitosis. Interphase is composed of three parts. The G1 period (gap 1 period) is a time of active synthesis of RNA, ribosomes, and other cytoplasmic constituents accompanied by considerable cell growth. This is followed by the S period (synthesis period) in which DNA is replicated and doubles in quantity. Finally, there is a second gap, the G2 period, when the cell prepares for mitosis, the M period, by activities such as the synthesis of special division proteins. The total length of the cycle differs considerably between microorganisms, usually due to variations in the length of G1. During mitosis, the genetic material duplicated during the S period is distributed equally to the two new nuclei so that each has a full set of genes. There are four phases in mitosis. In prophase, the chromosomes each with two chromatids become visible and move toward the equator of the cell. The mitotic spindle forms, the nucleolus disappears, and the nuclear envelope begins to dissolve. The chromosomes are arranged in the center of the spindle during metaphase and the nuclear envelope has disappeared. During anaphase the chromatids in each chromosome separate and move toward the opposite poles of the spindle. Finally during telophase the chromatids become less visible, the nucleolus reappears, and a nuclear envelope reassembles around each set of chromatids to form two new nuclei. For example, the nuclear envelope does not disappear in many fungi and some protozoa and algae. Frequently cytokinesis, the division of the parental cell's cytoplasm to form new cells, begins during anaphase and finishes by the end of telophase. However, mitosis can take place without cytokinesis to generate multinucleate or coenocytic cells. In mitosis the original number of chromosomes is the same after division and a diploid organism will remain diploid or 2n (i.e., it still has two copies of each chromosome). Frequently a microorganism reduces its chromosome number by half, from the diploid state to the haploid or 1n (a single copy of each chromosome).

Requirement : Onion root, Acetocarmine stain, glycerine, compound microscope etc.

Procedure : Onion root tips cut & boiled in a boiling tube having the acetocarmine stain solution in it for few minutes and meshed with the slide. After staining in glycerine it is observed for various stages in the compound microscope.

Observations and results : As observed in the compound microscope various stages of mitosis cell division.

PREPARATION OF NUTRIENT AGAR PLATES AND NUTRIENT BROTH

Requirements :

Nutrient Agar

Beef Extract	0.3%
Peptone	0.5%
Agar	1.5%

Procedure :

1. You will be making 200 ml of Nutrient Agar. To weigh out Beef Extract, first tare a tongue depressor, then dip it into the Beef Extract and weigh. Adjust the amount of Beef Extract until the correct amount is obtained. Be sure to be careful not to get Beef Extract on to the balance you need to weight out enough Beef Extract to get a 0.3% solution. Place the tongue depressor into the flask, beef extract side down.
2. Tare a weigh boat and weigh out enough Peptone and add that to the flask.
3. Add 200 ml of distilled water and swirl to dissolve the peptone and beef extract. Check the pH, it should be 7.0.
4. Tare a weigh boat and weigh out enough agar and add that to the flask.
5. With a Bunsen burner, tripod, asbestos wire-gauze, heats the medium to boiling to dissolve the agar. Keep the rotating the flasks to prevent the agar from cooking onto the bottom of the flask watch. As soon as it begins to boil take it off the heat and put it on to the bench. Allow it to cool a few minutes.
6. While the agar is still warm, but not hot, pipette 3 ml each into 413 × 100 mm screw cap culture tubes.
7. Label the flask and your tubes with your name.
8. After preparation of your medium, the instructor will take you to the autoclave.
9. Place your media in the autoclave. After discussion of the parts of the autoclave, autoclave the medium for 20 minutes.

Nitrate Broth

Beef extract	3 gm
Peptone	5 gm
Potassium nitrate	1 gm
Distilled water	1000 ml

Final pH 7.0 at 25°C

Procedure : Weigh and mix the chemical ingredients and autoclave at 15psi at 121°C.

Observations and results : As observed in the experiment

INOCULATION OF AGAR SLANTS, AGAR PLATE

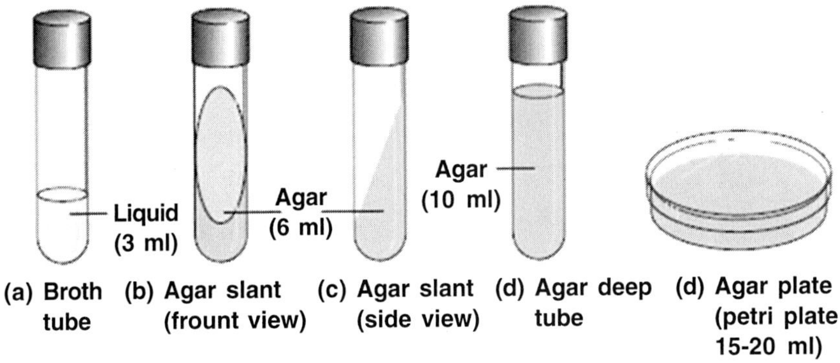

Fig. 3. Formation of Agar Plate.

Requirement : Agar slants are prepared to inoculate microbial culture. To prepare agar slant a required number of culture tubes are taken and about 12 to 15 ml of liquefied agar medium is poured in each of them.

The tubes are now cotton-plugged and sterilized in an autoclave. After the sterilization is over the tubes are taken out and are placed in slanting (sloping) position for sometimes, the tubes get cooled and the medium in them is solidified resulting in a slope surface.

Procedure :

1. The tube containing inoculum and the tube containing agar slant are held in the left hand and the inoculation loop/needle in the right hand. Tubes should almost be parallel to the ground to avoid contamination.
2. Both the tubes are opened by removing the cotton-plug with fingers of the right hand and the open mouth of the tubes is sterilized by passing through the flame twice.
3. Immediately after deplugging and sterilizing the mouth of the tubes, the loop/needle is also flame-sterilized and is inserted within the agar surface of the inoculum containing tube to quench the heat, and a small bit of inoculum is taken on the loop/needle tip.
4. The inoculum containing loop/needle is taken out and brought in within the agar slant containing tube where the inoculum is just rubbed on the surface of the agar slant.

5. Starting from plug removal from the mouth of the tubes to the rubbing of the inoculum on the surface of the agar slant should be taken quickly to avoid contamination.
6. When inoculation is complete, the open mouths of tubes and the cotton plugs are sterilized by flame and the cotton plugs are replaced.
7. The inoculated tube is incubated under suitable temperature to favour rapid growth of microorganisms.

Observations and results : As observed in the experiment

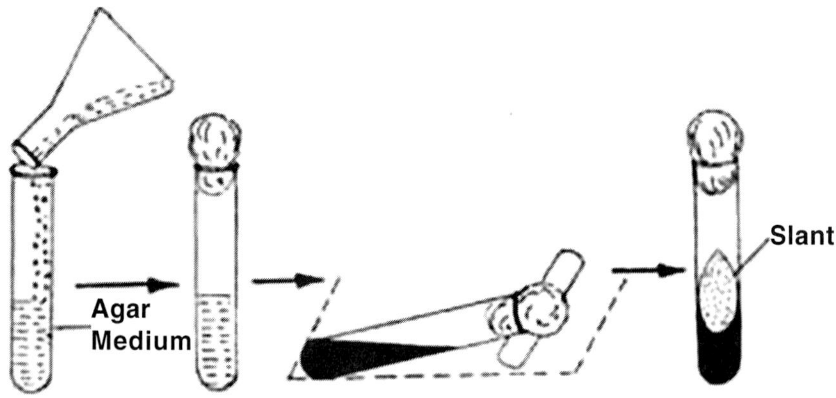

Fig. 4. Formation of Agar saint

CULTURE OF MICROORGANISMS USING VARIOUS TECHNIQUES

Requirement : Same as in the above experiment diagram as given above.

Procedure :

1. The main principle is to dilute the inoculum in successive tubes containing liquefied agar medium so as to permit a thorough distribution of bacterial cells within the medium.
2. The mixed culture of bacteria is diluted directly in tubes containing melted agar medium maintained in the liquid state at a temperature of 42 to 45°C (agar solidifies below 42°C).
3. The bacteria and the melted medium are mixed well. The contents of each tube are poured into separate Petri plates, allowed to solidify, and then incubated.

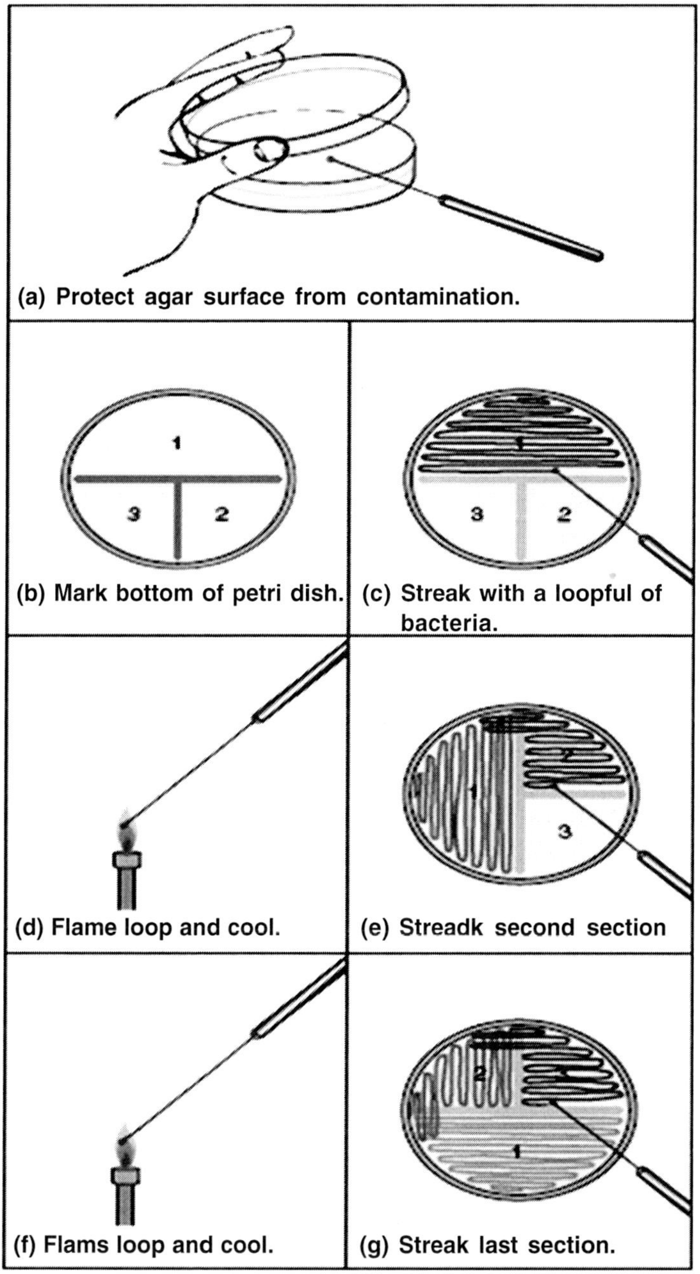

(a) Protect agar surface from contamination.

(b) Mark bottom of petri dish.

(c) Streak with a loopful of bacteria.

(d) Flame loop and cool.

(e) Streadk second section

(f) Flams loop and cool.

(g) Streak last section.

Fig. 5. Preparation of a streak plate.

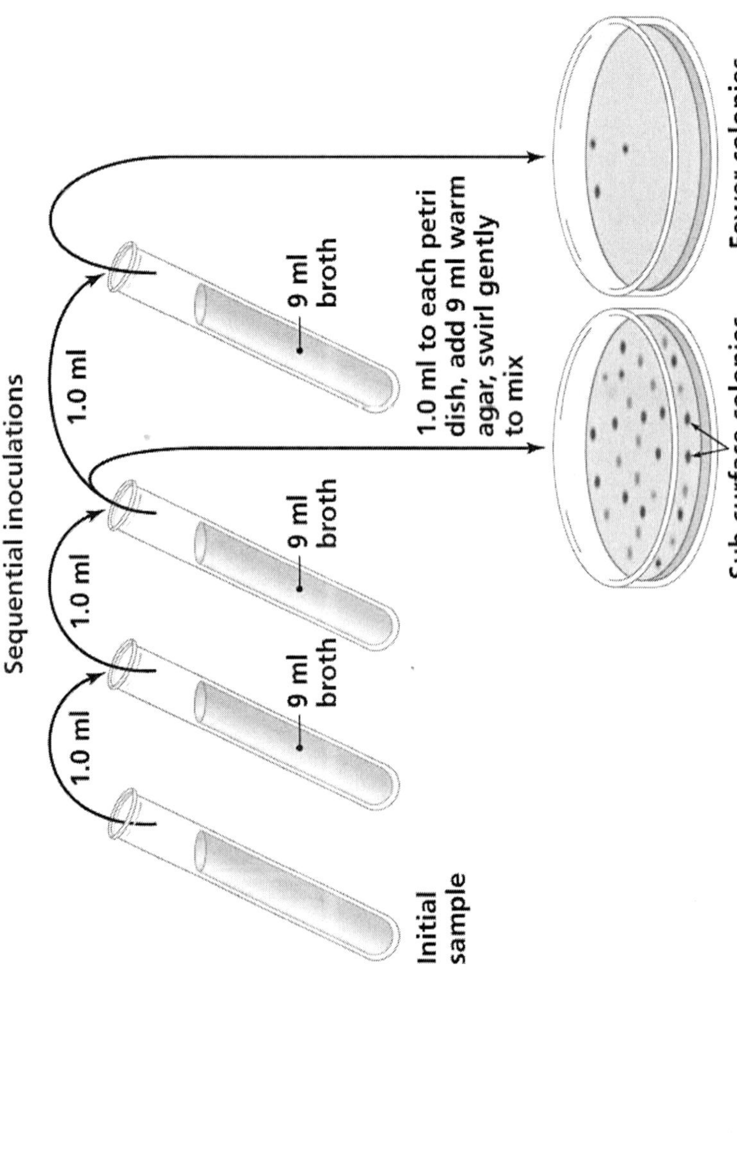

Fig. 6. The Pour-Plate Technique.

4. When bacterial colonies develop, one finds that isolated colonies develop both within the agar medium (subsurface colonies) and on the medium (surface colonies).
5. These isolated colonies are then picked up by inoculation loop and streaked onto another Petri plate to insure purity.

Observations and results : As observed in the experiment

SIMPLE AND DIFFERENTIAL STAINING PROCEDURES, ENDOSPORES STAINING, FLAGELLAR STAINING, CELL WALL STAINING, CAPSULAR STAINING, NEGATIVE STAINING

Materials : Staining bottles of the following: crystal violet, iodine, acetone/ alcohol or 95% alcohol, safranin.

Objectives :

1. To learn the Gram-stain procedure.
2. To learn to distinguish Gram-positive organisms from Gram-negative organisms.

Theory :

Gram Stain : The Gram stain is especially useful as one of the first procedures in identifying organisms because it reveals not only the morphology and the arrangement of the cells, but also information about the cell wall. Near the turn of the century, Christian Gram devised the staining procedure when trying to stain bacteria so that they contrasted with the tissue sections he was observing. Many years later, it was found that purple (Gram-positive) bacteria had thick cell walls of **peptidoglycan,** while pink (Gram-negative) bacteria had much thinner cell walls of peptidoglycan surrounded by an additional membrane. The thick cell wall retains the purple dye in the procedure, but the thin wall does not

Procedure for Gram Stain

1. Put two drops of water on a clean slide. In the first drop, make a suspension of the unknown organism to be stained just as you did for a simple stain.

2. In the second drop, mix together a known Gram-positive organism and a known Gram-negative organism. This mixture is a control to ensure that your Gram-stain procedure (Fig. 8) is giving the proper results. Heat-fix the slide. 2. Place the slide on a staining bar across a sink (or can). Alternatively, hold the slide with a clothespin or forceps over a sink.
3. Flood the slide with crystal violet until the slide is completely covered. Leave it on for 6–30 seconds and then discard into the sink. The timing is not critical. Rinse the slide with water from a wash bottle or gently running tap water. 4. Flood the slide with Gram's iodine for about 12–60 seconds and wash with water.
4. Hold the slide at a 45° angle and carefully drip acetone/ethanol over it until no more purple dye runs off. Immediately wash slide with tap water. Thicker smears may take longer than thinner ones, but acetone/alcohol should usually be added for 1–2 seconds and no more than 5 seconds. Timing is critical in this step.
5. Flood the slide with safranin and leave it on for 10–30 seconds timing is not important.
6. Wash with tap water. Safranin is a counter stain because it stains the cells that have lost the purple dye.
7. Blot the slide carefully with a paper towel to remove the water, but do not rub from side to side. When it is completely dry, observe the slide under the microscope. Remember that you must use the oil immersion lens to observe bacteria. Compare your stain to the control mixture on the same slide and with color plate.
8. 8. Describe the appearance of your stained bacteria in the Results section of the Laboratory Report.

Acid-fast Stain

The acid-fast stain is useful for identifying bacteria with a waxy lipid cell wall. Most of these organisms are members of a group of bacteria called mycobacteria. Although there are many harmless bacteria in this group, it also includes *Mycobacterium tuberculosis,* which is the cause of tuberculosis in humans. These organisms have a Gram-positive cellwall structure, but the lipid in the cell wall prevents staining with the Gram-stain dyes. In the Ziehl-Neelsen (Kinyoun modification) acid-fast stain procedure, the dye carbolfuchsin stains the waxy cell wall. Once the lipid-covered cell has been dycd, it cannot easily be decolorized even with alcohol containing HCL (which is called acid-alcohol). Nonmycobacteria are also dyed with the carbolfuchsin, but are decolorized by acid-alcohol. These colorless organisms are stained with methylene blue so they contrast with the pink acid-fast bacteria that were not decolorized. The reason this stain is important is that one of the initial ways tuberculosis can be diagnosed is by the presence of *Mycobacterium* in a patient's sputum. (Sputum is a substance that is coughed up from the lungs and contains puslike material.) Tuberculosis is a very serious disease worldwide and is now seen in the United States after decreasing for about 80 years. Since the process of finding acid-fast organisms in sputum is quite difficult and time-consuming, this test is usually performed in state health laboratories.

Procedure for Acid-Fast Stain (Kinyoun modification)

1. Prepare a smear of the material and heat-fix.
2. Cover the smear with Kinyoun carbolfuchsin and stain for to 3–5 minutes. Do not heat.
3. Rinse with water.
4. Decolorize with acid-alcohol for 10–30 seconds.
5. Rinse with water.
6. Counter stain with methylene blue for 20–30 seconds.
7. Rinse with water.
8. Blot dry carefully and examine under the oil immersion lens.
9. Record results.

Procedure for Capsule Stain

1. Make a suspension of the organism in a drop of water on a clean slide.
2. Put a drop of India ink next to it.
3. Carefully lower a cover slip over the two drops so that they mix together. There should be a gradient in the concentration of the ink.

4. Examine under the microscope and find a field where you can see the cells surrounded by a halo in a black background.
5. Drop slides in a beaker or can of boiling water and boil for a few minutes before cleaning. This is necessary because the bacteria are not killed in the staining process.
6. Record results.

Procedure for Endospore Stain

1. Prepare a smear on a clean slide and heat-fix.
2. Add about an inch of water to a beaker and bring it to a boil.
3. Place two short staining bars over the beaker and place a slide on them.
4. Tear a piece of paper towel a little smaller than the slide and lay on top of the smear. The paper prevents the dye from running off the slide.
5. Flood the slide with malachite green and steam for 5 minutes. Continue to add stain to prevent the dye from drying on the slide.
6. Decolorize with water for about 30 seconds by flooding with water or holding under gently

   Observations and results : As observed in the experiment

BACTERIAL COLONY COUNTING

Requirement : Glassware & bacterial sample.

Serial Dilution. Preparing a dilution in steps instead of one dilution.

Objectives

1. To enumerate bacteria using the plate count method.
2. To understand the use of dilutions.
3. To gain insight into the number of organisms that can be present in a clear liquid and a turbid liquid.

Procedure :

1. Label all water blanks with the dilution, as shown in figure.
2. Melt 4 TS agar deeps and hold at 50°C. It is very important not to let the deeps cool much lower than 50°C because the agar will harden and will have to be heated to boiling (100°C) before it will melt again (Fig. 7).

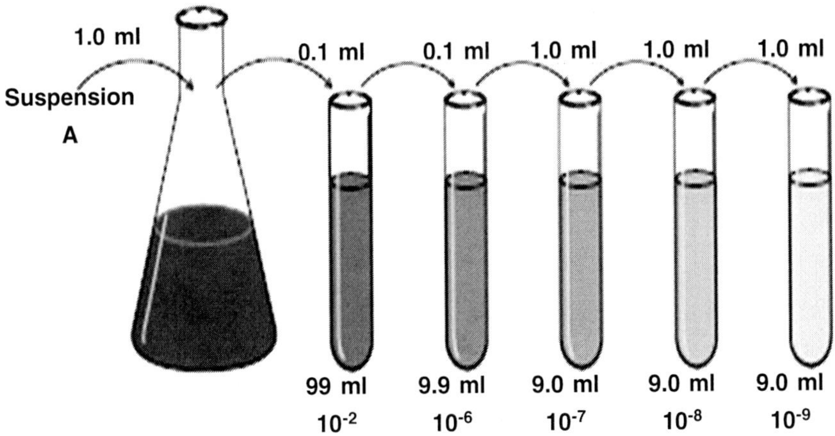

Fig. 7. The Pour-Plate Technique.

3. Make serial dilutions of the bacterial suspension.
   (a) Mix the bacterial suspension by rotating between the hands and transfer 1.0 ml of the suspension to the 99-ml water blank labeled 10:2. Discard the pipet.

   (b) Mix well and transfer 0.1 ml of the 10:2 dilution to the 9.9-ml water blank labeled 10:4. Discard pipette.

   (c) Mix well and transfer 0.1 ml of the 10:4 dilution to the 9.9-ml tube labeled 10:6.Discard pipette.

   (d) Mix and transfer 1.0 ml of the 10:6 dilutions to the 9.0-ml tube labeled 10:7. (Note: change from 0.1 ml to 1.0 ml.) Discard pipette.

   (e) Mix and transfer 1.0 ml of the 10:7 dilutions to the 9.0-ml tube labeled 10:8. Discard pipette.

   (f) Mix and transfer 1.0 ml of the 10:8 dilutions to the 9.0-ml tube labeled 10:9. Discard pipette.

4. Place samples of the dilutions 10:9, 10:8, 10:7, and 10:6 into sterile labeled Petri plates as follows:
   (a) Mix the 10:9 dilution and transfer 1.0 ml into a sterile Petri plate labeled 10:9. Add a tube of melted agar (wipe off the outside of

the tube before pouring) and swirl gently by moving the plate in a figure eight pattern on the bench. Do not discard pipette.

(b) Mix the 10:8 dilutions and with the same pipette transfer 1.0 ml to the Petri plate labeled 10:8. Add melted agar and mix.

(c) Mix the 10:7 dilutions again and using the same pipette transfer 1.0 ml to the Petri plate labeled 10:7. Add melted agar and mix.

(d) Mix the 10:6 dilutions and using the same pipette transfer 1.0 ml to the Petri plate labeled 10:6. Add melted agar and mix. Discard the pipette.

5. Invert the plates after you are sure the agar has hardened (about 5 minutes), and incubate at 37°C.

Observations and results : As observed in the experiment.

4 CELL BIOLOGY AND GENETICS

CELL CYCLES

The onion root tip and the whitefish blastula remain the standard introduction to the study of mitosis. The onion has easily observable chromosomes, and the whitefish has one of the clearest views of the spindle apparatus. The testis of the grasshopper and the developing zygote of the roundworm Ascaris are the traditional materials used for viewing the various stages of meiosis. In a single longitudinal section of a grasshopper testis, one can usually find all of the stages of meiotic development. The stages are also aligned from one pole of the testis. Few other meiotic samples are as convenient. For most materials, meiosis occurs in a more randomly distributed pattern throughout the testis.

Ascaris is utilized to observe the final stages of development in eggs (oogenesis). The Ascaris egg lies dormant until fertilized. It then completes meiosis, forming 2 polar bodies while the sperm nucleus awaits fusion with the female nucleus. When this phenomenon is coupled with the large abundance of eggs in the Ascaris body, it makes an ideal specimen for observing the events of fertilization, polar body formation, fusion of pronuclei, and the subsequent division of the cell (cytokinesis).

Interphase G1-S-G2

The stages of mitosis were originally detailed after careful analysis of fixed cells. More recently, time-lapse photography coupled with phase-contrast microscopy has allowed us to visualize the process in its entirety, revealing a dynamic state of flux. In early work, so much emphasis was placed on the movement of the chromosomes that the cell was considered to be "at rest" when not in mitosis. As significant as mitotic division is, it represents only a small fraction of the life span of a cell. Nonetheless, you may still come across the term "resting phase" in some older texts. This term is rarely used today, and the term interphase is sufficient for all activities between 2 mitotic divisions. The cell is highly active during interphase and most of the metabolic and genetic functions of the cell are reduced during the physical division of the nuclear and cellular materials (mitosis). Interphase is divided into 3 sub-phases, G1, S, and G2. The basis for this division is the synthesis of DNA. While the entire cycle may be as long as 24 hours, mitosis is normally less than 1 hour in length.

Because of the synthesis of DNA in interphase, the amount of DNA per nucleus is different depending on which subphase of interphase the cell is in DNA can be measured using the fuelgen reaction and a microspectrophotometer. The basic amount of DNA in a haploid nucleus is given the value C. A diploid nucleus would be 2C. A triploid and tetraploid cell would be 3C and 4C, respectively.

However, when nuclei are actually measured, diploid cells in Interphase can be divided into 3 groups; some are 2C and some are 4C, while a few are at intermediate values between 2C and 4C. The conclusion is that the genetic material (DNA) and presumably the chromosomes must duplicate. The period within interphase and during which DNA is synthesized is termed the S phase (for synthesis). The period of interphase preceding the S phase is the G1 phase (for 1st-Growth Phase), while the period subsequent to the S phase is the G2 phase (for 2nd-Growth Phase). During the G1 period, the cell is generally increasing in size and protein content. During S, the cell replicates the chromosomes and synthesizes DNA. During G2, it continues to increase in size, but also begins to build a significant pool of ATP and other high-energy phosphates, which are believed to be a significant part of the triggering mechanism for the subsequent karyokinetic and cytokinetic events of mitosis. Mitosis returns the cells to the 2C state. Meiosis reduces the amount of DNA even further, to 1C. Meanwhile, the number of chromosomes (designated with the letter N) is also changing. For a diploid cell, the number of chromosomes is twice that of a haploid, or 2N. During mitosis, a diploid cell would go from one 2N cell to two 2N cells. Since the daughter cells have the same chromosome number as the parent, mitosis is also referred to as equational division. If a diploid (2N) cell undergoes meiosis, it will result in 4 haploid cells, each 1N. Thus, meiosis is also referred to as reductional division. Refer to the figure below for comparison of C and N values during division.

It is possible to visualize the process of DNA synthesis within either nuclei or chromosomes by the incorporation of a radioactive precursor to DNA into cells and subsequent detection by autoradiography. Incorporation of thymidine, a DNA precursor, will only occur during the S phase, and neither during G1 nor G2. If a pulse (short period of exposure) of 3H-thymidine is presented to cells, those that are in the S-period will incorporate this radioactive substance, while all others will not. Careful application of the pulse will allow the calculation of the duration of the S-phase. By knowing the timing for the entire cycle (from mitosis to mitosis), one can deduce the G1 and G2 periods. Meiotic division differs from mitosis in that there are 2 division cycles instead of 1. In the first cycle, interphase is the same as for mitosis. That is, there is an S-phase with corresponding G1 and G2. During the second interphase, however, no DNA synthesis occurs. Consequently, there is no G1 or G2 in the second interphase. The result is that chromosomes are replicated prior to meiosis, and do not replicate again during meiosis.

Prophase

The first phase of mitosis is marked by the early condensation of the chromosomes into visible structures. At first, the chromatids are barely visible, but as they continue to coil, the chromosomes become thicker and shorter. The nuclear envelope is still present during this stage, as are any nucleolar structures. The centrioles are moving to the poles of the cell and spindle fibers are just beginning to form.

Metaphase

During the middle phase of karyokinesis, the chromosomes line up in the center of the cell, and form a metaphase plate. Viewed on edge, the chromosomes appear to be aligned across the entire cell, but viewed from 90° they appear to be spread throughout the entire cell (visualize a plate from its edge or from above). Each chromosome has a clear primary constriction, the centromere, and attached to each is a definitive spindle fiber. The spindle apparatus is completely formed, and the centrioles have reached their respective poles. The nucleolus and the nuclear envelope have disappeared.

Anaphase

The movement phase begins precisely as the 2 halves of a chromosome, the chromatids, separate and begin moving to the opposite poles. The centromere will lead the way in this process, and the chromatids form a V with the centromeres pointing toward the respective poles.

Telophase

The last phase is identified by the aggregation of the chromatids (now known as chromosomes) at the respective poles. During this phase, the chromosomes uncoil, the nuclear envelope is resynthesized, the spindle apparatus is dismantled, and the nucleolus begins to appear.

Meiosis

For meiosis, the phases prophase, metaphase, anaphase, and telophase are identified, but because there are 2 divisions, there are 2 sets. These are designated by Roman numerals; thus, prophase I, metaphase I, anaphase I, telophase I, interphase, prophase II, metaphase II, anaphase II, and telophase II. Interphase is normally not designated with a Roman numeral. Because of the significance of the chromosome pairing that occurs in prophase I, it is further subdivided into stages.

The phases of prophase I are named for the appearance of a thread like structure, known as nema. Leptonema means "thin thread" and leptotene is the adjective applied to the term stage, i.e., proper terminology is the "leptotene stage" of prophase I. The word stage is often omitted.

Prophase I: Leptotene

This stage is marked by the first appearance of the chromosomes when they are in their most extended form (except for during interphase). They appear as a string with beads. The beads are known as chromomeres. The chromatids have already replicated prior to this phase, but typically, the replicated chromatids cannot be observed during the leptotene stage.

Prophase I: Zygotene

Zygos means "yoked," and during this stage, the homologous chromosomes are seen as paired units. The chromosomes are shorter and thicker than in leptotene, and in some cells they remain attached to the nuclear envelope at the points near the aster. This gives rise to an image termed the "bouquet." This attachment is rare in invertebrates and absent in plants, where the chromosomes appear to be a tangled mass.

Prophase I: Pachytene

When the pairing of zygotenes is complete, the chromosomes appear as thick strings, or pachynema. The chromosomes are about ¼ the length they were in leptotene, and there are obviously 2 chromosomes, with 2 chromatids in each bundle. The 2 chromosomes are referred to as a "bivalent," while the same structure viewed as 4 chromatids is known as a "tetrad."

Prophase I: Diplotene

This stage results as the gap between the 2 homologous chromosomes widens. The homologs have already paired during zygotene, recombined during pachytene, and are now beginning to repel each other. During this stage, the chromosomes of some species uncoil somewhat, reversing the normal direction typical of prophase. As the chromosomes separate, they are observed to remain attached at points known as "chiasmata." These are believed to be the locations where genetic recombination of the genes has taken place.

Prophase I: Diakinesis

Prophase I end as the homologs completely repel each other. The chromosomes will continue to coil tightly (reversing the slight uncoiling of the diplotene) and will reach their greatest state of contraction. As diakinesis progresses, chiasmata appear to move toward the ends of the chromosomes, a process known as "terminalization." Since this stage is the end of prophase, the nucleolus usually disappears, along with the nuclear envelope.

Metaphase I

The tetrads move toward the center and line up on a metaphase plate. The nuclear envelope completely disappears. As the tetrads align themselves in the middle of the cell, they attach to spindle fibers in a unique manner. The centromeres of a given homolog will attach to the spindles from only 1 pole.

Anaphase I

The unique event occurring at this phase is the separation of the homologs. In contrast to mitotic anaphase, the centromeres of a given homolog do not divide, and consequently each homolog moves toward opposite poles. This results in a halving of the number of chromosomes, and is the basis of the reduction division that characterizes meiosis.

Telophase I: Interphase

Telophase in meiosis is similar to that of mitosis, except that in many species, the chromosomes do not completely uncoil. If the chromosomes do uncoil and enter a brief interphase, there is no replication of the chromatids. Remember that the chromatids have already been replicated prior to prophase I.

Prophase II: Telophase II

These phases are essentially identical in meiotic and mitotic division: the only distinction is that the chromosome number is half of that prior to meiosis. Each chromosome (homolog) is composed of 2 chromatids, and during anaphase II, the 2 chromatids of each chromosome move apart and become separate chromosomes. While the 2 chromatids remain attached at the centromere, they are known as chromatids. Immediately upon separating, each chromatid becomes known as a chromosome and is no longer referred to as a chromatid. This is the reason that a cell can divide 1 chromosome (with 2 chromatids) into 2 cells, each with a chromosome the term applied to the chromatid is changed.

Damage Induced During Division

In 1949, Albert Levan developed what was to become known as the Allium test for chromosome damage. Growing roots from onion bulbs were soaked in various agents and analyzed for their effect on mitosis. It was discovered that caffeine, for example, caused complete inhibition of mitosis, primarily through the inhibition of cell plate formation. This test was later used extensively by B.A. Kihlman and extended to other higher plants. Kihlman found that 1 to 24 hour treatments of cells with caffeine and related oxypurines not only inhibited mitosis, but induced significant chromosome alterations (aberrations). Specifically, this treatment induced "stickiness" and "pseudochiasmata." Stickiness is the clumping of chromosomes at metaphase and the formation of chromatin bridges at anaphase. Pseudochiasmata is the formation of side-arm bridges during anaphase. Caffeine also causes the formation of other chromosome and chromatid breaks and exchanges. Colchicine, a drug that inhibits spindle fiber formation during mitosis, can be added to the growing cells to halt cell division at metaphase. This often will result in a doubling of the chromosome number, since colchicine typically inhibits cytokinesis, but not karyokinesis. The doubled chromosomes will fuse within a single nucleus, thus increasing the ploidy value of the nucleus. Moreover, methylated oxypurines (caffeine, theophylline, and 8-ethoxycaffeine) are inhibitors of cell plate formation. Treatment with these agents for 0.5 to 1 hour, with concentrations as low as 0.02% to 0.04%, results in the cell's failure to undergo cytokinesis; in addition, the nuclei do not fuse into a single unit. Thus, treatment with any of these agents should result in binucleate or multinucleate cells.

In addition, alkylating agents such as (di-chloroethyl)methylamine or nitrogen mustard, di-epoxypropyl ether (DEPE) and b-Propiolactone (BPL), nitrosocompounds (N-Nitroso-N-methylurethan (NMU)), N-Methyl phenylnitrosamine (MPNA), N-hydroxylphenylnitrosamine-ammonium (cupferron), and 1-Methyl-3-nitrosoguanidine (MNNG) have all been indicated as potent chromosome-breaking agents. Other compounds have included such things as maleic hydrazide, potassium cyanide, hydroxylamine, and dyes such as acridine orange in visible light. The damages involve abnormal metaphases, isochromatid breaks, chromatid exchanges, and anaphase bridges, to name a few.

MEIOSIS IN FLOWER BUDS OF *ALLIUM CEPA* (ACETOCARMINE STAIN)

Materials

1. Onion flower
2. Acetocarmine stain
3. Glass slides
4. Cover slips
5. Blotting paper

Procedure

Select appropriate flower buds of different size from the inflorescence.

1. Fix them in Carnoy's fluid, which is used as fixative.
2. Take a preserved flower bud and place it on a glass slide.
3. Separate the anthers and discard the other parts of the bud.
4. Put 1 or 2 drops of acetocarmine stain and squash the anthers.
5. Leave the material in the stain for 5 minutes.
6. Place a cover slip over them and tap it gently with a needle or pencil.
7. Warm it slightly over the flame of a spirit lamp.
8. Put a piece of blotting paper on the cover slip and apply uniform pressure with your thumb.
9. Observe the slide under the microscope for different meiotic stages.

Interphase

Before undergoing in meiosis-I, each cell will remain in an interphase, during which the genetic materials are duplicated due to active DNA replication.

Prophase-I

In the first meiotic division, production in the chromosome number occurs without separation of chromatids. Prophase is the longest phase and has 5 stages.

Leptotene

Chromosomes appear as long threadlike structure interwoven together. Chromosomes display a beaded appearance and are called chromomeres. Ends of chromosomes are drawn toward nuclear membrane near the centriole. In some plants, chromosomes may form synthetic knots.

Zygotene

The homologous chromosomes pair with one another, gene by gene, over the entire length of the chromosomes. The pairing of the homologous chromosomes is called synapsis. Each pair of homologous chromosomes is known as bivalent.

Pachytene

Each paired chromosomes become shorter and thicker than in earlier substages and splits into 2 sister chromatids except at the region of the centromere. As a result of the longitudinal division of each homologous chromosome into 2 chromatids, there are 4 group of chromatids in the nucleus parallel to each other, called tetrads.

Diplotene

During the diplotene stage, chiasmata appear to move towards the ends of the synapsed chromosomes in the process of terminalization. Repulsion of homologous chiasmata are very clear in pachytene because of the increased condensation of the chromosomes.

Diakinesis

The chromosomes begin to coil, and so become shorter and thicker. Terminalization is completed. The nucleolus detaches from the nucleolar organizer and disappears completely. The nuclei envelope starts to degenerate and spindle formation is well underway.

Metaphase-I

The bivalents orient themselves at random on the equatorial plate. The centromere of each chromosome of a terminalized tetrad is directed toward the opposite poles. The chromosomal microtubular spindle fibers remain attached, with the centromeres and homologous chromosomes ready to separate.

Anaphase-I

It is characterized by the separation of whole chromosomes of each homologous pair (tetrad), so that each pole of the dividing cell receives either a paternal or maternal longitudinally double chromosome of each tetrad. This ensures a change in chromosome number from diploid to monoploid or haploid in the resultant reorganized daughter nuclei.

Telophase-I

The chromosomes may persist for a time in the condensed state, the nucleolus and nuclear membrane may be reconstituted, and cytokinesis may also occur to produce 2 haploid cells.

Metaphase-II

Metaphase-II is of very short duration. The chromosomes rearrange in the equatorial plate. The centromere lies in the equator, while the arms are directed toward the poles. The centromeres divide and separate into 2 daughter chromosomes.

Anaphase-II

Daughter chromosomes start migrating toward the opposite poles and the movement is brought about by the action of spindle fibers.

Telophase-II

The chromosomes uncoil after reaching the opposite poles and become less distinct. The nuclear membrane and nucleolus reappear, resulting in the formation of 4 daughter nuclei, which are haploid.

Cytokinesis

This separates each nucleus from the others. The cell wall is formed and 4haploid cells are produced.

DIFFERENTIAL STAINING OF BLOOD

To identify different stages of white blood cells in human blood.

Introduction

White blood corpuscles (WBCs) or leucocytes are colorless, actively motile, nucleated living cells. They are variable in size and shape and exhibit characteristic amoeboid movement. Under certain physiological and pathological conditions, WBCs can come out of the blood vessels by a process called diapedesis. The lifespan of WBCs is 12–15 days. Unlike red blood corpuscles, leucocytes or WBCs have nuclei and do not contain hemoglobin. They are broadly classified into 2 groups:

1. Granulocytes and
2. Agranulocytes.

Granulocytes develop from red bone marrow and agranulocytes from lymphoid and myeloid tissues. Granulocytes are subdivided into 3 groups:

a. Eosinophils
b. Basophils and
c. Neutrophils,

Agranulocytes are subdivided into 2 groups:

a. Lymphocytes and
b. Monocytes.

The main features of WBCs are:

- Phagocytosis
- Antibody formation and
- Formation and secretion of lysozyme, heparin, etc.

Principle

WBCs are the type of blood cells that have a nucleus but no pigment. They are important in defending the body against diseases because they produce antibodies against any foreign particle or antigens. WBC can be divided into 2 types granulocytes and agranulocytes.

1. **Granulocytes** consists of neutrophils, eosinophils, and basophils.
    a. **Neutrophils** Do not stain with either acidic or basic dyes. They have manylobed nucleus and are called polymorphonuclear leucocytes or polymorphs. They constitute about 76.9% of the total leucocytes found.
    b. **Eosinophils** A lobed nucleus is present, with cytoplasmic granules that stain with acidic dyes. They constitute 1%–4% of the total leucocyte count.
    c. **Basophils** It has a lobed nucleus and the cytoplasm contains granules, which stain with basic dyes. It comprises 0% to 4% of the total leucocyte count.
2. **Agranulocytes** Consist of lymphocytes and monocytes.
    a. **Lymphocytes** This is a type of WBC with a very large nucleus. It is rich in DNA and a small amount of clear cytoplasm is present.

Procedure

1. The fingertip is cleaned with cotton dipped in alcohol and pricked with a sterilized needle.
2. One or two drops of blood are placed on the right side of the slide, and with the help of another clean slide, the blood is drawn so that a thin smear is formed and dried.
3. The dried smear is fixed using zinc acetone, 3-methyl alcohol, or absolute alcohol for 5 min.

4. The smear is dried and stained with Giemsa or Leishman stain. Subsequently, the distilled water (double the amount of water of the stain) is added onto the same smear slide.
5. The smear is mixed using a pipette for 10 to 15 min.
6. The slides are kept in running water to remove excess stain, and then dried.
7. The slide is observed under an oil immersion lens.

BUCCAL EPITHELIAL SMEAR AND BARR BODY

The study of the Barr body from the (female) smear of Buccal epithelial cells.

Materials

1. Buccal epithelial cells
2. Giemsa stain
3. Carnoy's fixative
4. Slides
5. Cover slip
6. Microscope, etc.

Procedure

1. Gently rub the inside of the cheek with a flat rounded piece of wood and transfer the scraping over a clean glass slide.
2. Then, made a thin film of cells on the slide and keep them for air-drying.
3. Air-dried smear was kept in Carnoy's fixative for 30–35 minutes.
4. Then, the Giemsa stain was poured and allowed to stand for 20–25 minutes.
5. After staining, the slide was washed with distilled water to remove the excess stain.
6. Finally, the slide was kept for air-drying and then observed under the microscope.

Observations

We found that very lightly stained cells are scattered here and there in thesmear. In the cells, violet-Barr bodies are observed inside a pink nucleus. A Barr body is nothing but an inactivated (heterochromatinized) X chromosome. It was first observed by Murray Barr in 1949. It is found only in female cells, because in those 1 X chromosome is enough for metabolic activity. It is absent in male somatic cells, because there only 1 X chromosome is present, which is in an active state.

Precautions

The smear or film should be uniform and thin over the glass slide so that the cells will not overlap each other.

STUDY OF THE MALARIA PARASITE

It is an intracellular parasite found in the blood of men. They cause malaria, are malarial parasites, and have 2 hosts.

a. Man-Primary host

b. Female Anopheles-Secondary host.

The life cycle in men is sexual and in mosquitoes is asexual. There is an alteration of generation of asexual and sexual cycles. Malarial parasites are transmitted from person to person by the adult female Anopheles mosquito. The male mosquito does not play any role in the transmission of malaria, because they do not feed on blood.

Life Cycle of Plasmodium

Includes 3 stages :

1. Pre-erythrocytic or exoerythrocytic cycle in humans
2. Erythrocytic or schizygotic cycle in RBC
3. Sexual or gametogenic in female mosquitoes.

1. In Man : When the female Anopheles mosquito bites a man and infects the plasmodium into his blood, this infection stage of plasmodium is called sporozoire.

2. **Exoerythrocytic Cycle :** The exoerythrocytic cycle is the life cycle of parasites inside the RBC of the host. This is an asexual phase, which results in the production of gametocytes. This cycle starts with the entry of merozoire into the RBC. In the RBC, the parasite enters, resulting in a resting period. It attains a round shape called a trophozoire. In the trophozoire, a large vacuole develops and pushes the nucleus to one side. This stage is known as the signet ring. The vacuole disappears; the parasite fills the RBC to become a schizont ring. The schizont mature and undergo fusion to form mesozoites. They are released into the blood stream by the bursting of RBC. The mesozoites attach fresh RBC and the cycle repeats. After a few generations, some of the mesozoites develop into gametocytes in the RBC. There are 2 types of gametocytes. One is the macrogametocyte, which has a small nucleus, large cytoplasm, and is circular. Both the gametocytes do not undergo further development until they reach the stomach of the Anopheles. If they cannot reach it, they disintegrate. When an Anopheles mosquito bites a malarial parasite patient for blood in the stomach of the mosquito, only mature gametocytes survive to develop into gametes. Others disintegrate in the process of gamete formation, called gametogamy.

Gametogamy

During gametogamy, the microgamete becomes active. It produces 6–8 slender nucleated bodies, called male gametes, by a process of exflagellation. The microgamete settles freshly in the stomach of the mosquito. The macrogametocyte undergoes the maturation phase and develops into a female.

Sporogony

The male gametes fuse with the female gamete to form a spherical zygote and remain inactive for some time. Later it transforms into an elongated wormlike mouth structure called ookinite. Ookinite pierces through the wall of the stomach and binds to the outer layer of the wall. There it becomes round, secretes a cyst wall, and grows in size. This stage is called oocyst. The nucleus divides into bits, each of which develops into slender sickle-shaped cell bodies called sporozoire. The mature cyst ruptures to liberate the sporozoites into the body cavity of the mosquito. Formation of sporozoites from zygotes is called sporogony. These sporozoites are ready to reach the salivary glands, and when a mosquito bites a healthy person, sporozoites are released to his blood stream, and the cycle repeats.

STUDY OF MENDELIAN TRAITS

Survey of Human Heredity

Some variations in human beings are inheritable. These variations can be studied using genetic principles. Some examples are given below:

Eye Color

Eye color is an example of Mendelian inheritance in man. Eye color is a polygenic trait and is determined primarily by the amount and type of pigments present in the eye's iris. Humans and animals have many phenotypic variations in eye color. In humans, these variations in color are attributed to varying ratios of eumelanin produced by melanocytes in the iris. The brightly colored eyes of many bird species are largely determined by other pigments, such as pteridines, purines, and carotenoids. Three main elements within the iris contribute to its color: the melanin content of the iris pigment epithelium, the melanin content within the iris stroma, and the cellular density of the iris stroma. In eyes of all colors, the iris pigment epithelium contains the black pigment, eumelanin. Color variations among different irises are typically attributed to the melanin content within the iris stroma. The density of cells within the stroma affects how much light is absorbed by the underlying pigment epithelium. Darker colors like brown are dominant over blue and gray. In some albinos, the iris is pink because the blood in the retinal layer is visible. The black circle in the center of the iris is the result of a sex-linked recessive gene.

Ears

Variations in the size and form of the ears and their position on the head indicate multiple gene inheritances. A few of them correspond to variations in a single gene.

Free Ear Lobes

Free ear lobes are dominant over attached ear lobes, but there is variation in those that are not attached. On the outer margin of the pinna is a rolled rim; there is a variation in its size. In some, it is almost lacking.m Nearly all persons have an enlarged portion of the cartilage that projects inwards from the rolled rim at a distinct point. It is called *Darwin's point,* which is inherited as dominant. In some persons, it is expressed only 1 ear. The inheritance of natural ear lobes is the expression of an autosomal dominant gene. Many persons show this characteristic expressed on only 1 side.

Widow's Peak

A widow's peak is a descending V-shaped point in the middle of the hairline (above the forehead). The trait is inherited genetically and dominant. This is one of the most important Mendelian traits found in human beings.

Tongue

The shape and size of the tongue is the result of many different genes. Some people have the ability to roll their tongue into a "U" shape. This is the expression of a dominant gene. The gene gives a few individuals the ability to fold the tongue.

Hitchhikers Thumb

The ability to hyperextend the thumb (extended backwards at the last joint) is due to a recessive allele (n), and a straight thumb (N) is dominant to hitchhiker's thumb.

PTC Tasting

The ability to taste phenylthio carbamide (PTC) is dominant (T) to the inability to taste it (t). Your instructor will provide you with small pieces of paper that have been previously soaked in this harmless chemical.

Hypertrichosis

This characteristic is controlled by helandric genes. It is characterized by hair on the pinna. It follows the linear pattern of inheritance that is transferred from father to son, but never to daughter.

ESTIMATION OF NUMBER OF ERYTHROCYTES (RBC) IN HUMAN BLOOD

Principle

To count the erythrocyte cells, or RBC, diluting fluid was used. Diluting fluid for RBC is an isotonic solution. This solution keeps the RBC cells unhemolyzed. The RBC was counted using an improved neubauer chamber, which has an area of 1 square millimeter and depth of 0.1 mm. It is made up of 400 boxes. The cells in the 4 corner boxes and 1 box in the middle were counted.

Reagents Required

EDTA : Ethylene diamine tetra acetic acid is used as an anticoagulant. RBC diluting solution: This is prepared by dissolving 0.425 gms of NaCl in 50 mL of distilled water.

Materials

1. Syringe
2. Needle
3. Alcohol
4. Vials
5. RBC pipette
6. Cotton
7. Neubauer counting chamber
8. Cover glass
9. Microscope

Procedure

Using a syringe, venous blood was drawn and poured into vials containing anticoagulant (EDTA) and mixed well. With the help of RBC, pipette blood was drawn up to the 0.5 mark. The tip of the pipette was cleaned with cotton and RBC diluting fluid was drawn up to the 101 mark. The pipette with blood and fluid was shaken well, avoiding bubbling, and kept aside for 5 minutes. The counting chamber was charged using the posture pipette. After charging the chamber, the cells were allowed to settle down to the bottom of the chamber. The chamber was placed on the stage of the microscope and using a 45X objective, the RBC cells in the smallest square were counted.

Comment

The normal range of RBC in an adult male is 5 to 6 million and in a female it is 4 to 5.05 million/mm3. In the case of polycythemia, the number of RBC increases. This disease is associated with heart disease. A low count of RBC results in anemia. Polycythemia may be pathologic due to the tumor in bone marrow, and the number of red cells may reach 11 million per cubic mm of blood. The present determination of total RBC count shows about ____/mm3 of blood, which is nearly equal to normal range.

Calculation

Number of cells in 1 mm^3 of blood

$$= \frac{\text{Number of cells counted} \times \text{dilution factor} \times \text{depth factor}}{\text{Area of chamber counted}}$$

$$= \frac{\text{Number of cells counted} \times 200 \times 10}{1/5}$$

$$= \text{Number of cells counted} \times 10{,}000$$

CULTURING AND STAINING OF E. COLI (GRAM'S STAINING)

E.coli colonies were isolated from water and inoculated in nutrient agar.

Composition of Nutrient Agar

1.	Peptones	5 gm
2.	Beef extract	3 gm
3.	NaCl	5 gm
4.	Agar agar	20 gm
5.	Distilled water	300 ml
6.	pH	(7.0 ± 0.2)

The above medium is prepared and is autoclaved at 15 lbs. The solution is poured into clean, sterilized petri dishes. After cooling, the petri dishes are kept in the refrigerator for 10 to 20 hrs. Then, the petridish is kept for incubation for 2 to 4 hrs for 37°C. The isolated *E.coli* inoculum is treated with the nutrient agar plate. The plates are inverted and incubated at 37°C for 24 hrs. After 24 hrs, whitish streaks indicate growth of *E.coli.* A portion of the colony is stained and mounted.

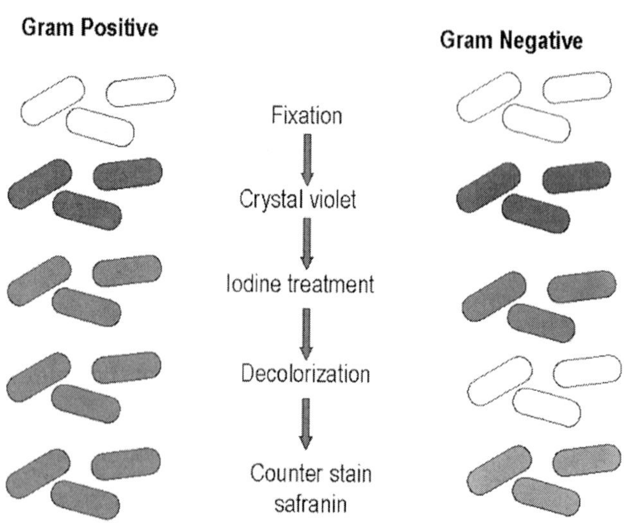

Fig. 8. Gram's Staining

Staining of E.coli

Objective : To stain bacteria (*E.coli*) using the Gram-staining method.

Materials

1. *E.coli* culture
2. Crystal violet
3. Gram iodine solution
4. Ethyl alcohol (98%)
5. Saffranine
6. Staining tray
7. Dropper
8. Glass slide
9. Microscope

Procedure

1. Prepare a smear of the culture on a glass slide and heat fix it.
2. Flood the smear with crystal violet for 1 minute.
3. The rinses are washed under slow running water and flood the smear with iodine solution. Let it stand for 1 minute.
4. Rinse the slide under slow running water and wash with 98% ethyl alcohol. Let it stand for 1 minute.
5. Rinse the slide under slow running water and add a few drops of safranin to the smear and keep it for 30 seconds.
6 Wash the smear and dry the slide.
7. Use a cover slip. Put a drop of cedar wood oil over the cover slip and observe under an oil immersion lens.

Result

If the material is Gram-negative, it is decolorized by alcohol and stained by safranin. If it is positive, it is stained by Gram iodine stain.

ABO BLOOD GROUPING AND Rh FACTOR IN HUMANS

Introduction

The ABO system of blood group was introduced by Landsteiner in 1900. He found 2 types of antigens present on the RBCs. They are antigen

A and antigen B. Similarly, there are 2 types of antibodies present in the plasma called antibody A and B. Based on the presence or absence of antigen and antibodies, human blood is classified into 4 groups A, B, AB, and O. The A group contains antigen A and antibody B. The B group contains antigen B and antibody A. The AB group contains both antigens A and B and no antibodies. The O group contains no antigen and both antibodies A and B. The ABO blood group is inherited by a set of multiple alleles. Presence of a particular factor is denoted by Rh factor discovered by Weiner from the rabbits immunized with the blood of the *Macaca rhesus* monkey.

DETERMINATION OF BLOOD GROUP AND Rh FACTOR

Materials

1. Blood sample
2. Applicator sticks
3. Antisera A,B, and Rh
4. Glass slide
5. Cotton
6. Spirit and disposable needles

Procedure

1. Clean a glass slide thoroughly.
2. Place a drop of antiserum A on the left side, antiserum B on the right side, and antiserum D at the center of the slide.
3. Clean the tip of the index finger with cotton soaked in spirit.
4. Prick the finger tip with the help of a sharp, sterilized disposable needle.
5. Place 3 drops of blood near the antisera.
6. Mix the blood and antiserum using application sticks.

Result

Observe your experimental result.

Conclusion

Draw conclusions based on your result.

5 Molecular Biology and Genetic Engineering

THE CENTRAL DOGMA

Introduction

The central dogma of modern biology is the conversion of the genetic message in DNA to a functional mRNA (transcription) and subsequent conversion of the copied genotype to a phenotype in the form of proteins.

The process of conversion of mRNA to a functional protein is known as translation. It involves the attachment of a messenger RNA to the smaller subunit of a ribosome, the addition of the larger subunit, plus initiation by a host of other factors. The entire process can be accomplished in the absence of a cell, if all of the necessary factors are present. Unfortunately, studies on translation and post-translational changes in protein structure are rather complex. They require a heavy investment of time and equipment. To some extent, the electrophoretic identification of proteins is part of this process, and the appearance of specific proteins can be monitored during any of the developmental processes.

To study the process of translation in any meaningful way requires that reasonably purified sources of mRNA, ribosomes, and amino acids be available. In addition, there is a requirement for various factors responsible for peptide chain initiation on the ribosome.

It is definitely an advanced technique, and requires mastery of many of the techniques presented in previous chapters. It must be performed on an independent basis, since the extensive time commitment does not lend itself to typical laboratory periods.

NUCLEIC ACIDS

Introduction

DNA is a high molecular weight macromoleculc. It is suitable for digestion with restriction enzymes. it is very important for the studies of molecular biology. Several methods are used for the isolation of genomic DNA from tissues of organisms .But isolation of quality DNA from plants (mainly alkaloid yielding plants) is rather difficult . Due to low yield of DNA high amount of plant tissues are required; hence old methods of DNA isolations aren't preferred. They are expensive and time consuming

In molecular terms, the process involves regulation using DNA as a primer molecule for the selective synthesis of RNA. DNA-primed RNA synthesis is termed "transcription."

Transcription is a complex series of reactions which involve the use of an RNA polymerase enzyme, known as transcriptase. If the reaction occurs in reverse, that is, with an RNA primer synthesizing DNA, the process is known as reverse transcription, and the enzyme is reverse transcriptase. This latter process is important for RNA virus replication in general, and is most significant when examining oncornagenic virus. Oncornagenic viruses are RNA-containing viruses that are also causal agents for some forms of cancer (primarily in birds). Reverse transcription is also important in the development of amphibians, and in the process of gene amplification.

TO ISOLATE GENOMIC DNA

REQUIREMENT : Liquid nitrogen, centrifuge machine, water bath, eppendorf tube, Micropipette, microtips, mortar-pestle glasswares, and chemicals as per the requirements and also, plant sample.

Principle : DNA isolation is a routine procedure to collect DNA for subsequent molecular or forensic analysis. There are three basic and one optional step in a DNA extraction:

1. Breaking the cells open commonly referred to as cell disruption or cell lysis, to expose the DNA within. This is commonly achieved by grinding or sonicating the sample.
2. Removing membrane lipids by adding a detergent.
3. Removing proteins by adding a protease (optional but almost always done).
4. Precipitating the DNA with an alcohol, usually ice-cold ethanol or isopropanol. Since DNA is insoluble in these alcohols, it will aggregate together, giving a pellet upon centrifugation. This step also removes alcohol-soluble salt.

Cellular and histone proteins bound to the DNA can be removed either by adding a protease or by having precipitated the proteins with sodium or ammonium acetate, or extracted them with a phenol-chloroform mixture prior to the DNA-precipitation. If desired, the DNA can be resolubilized in a slightly alkaline buffer or distilled water.

Materials

1. CTAB (Cetyl trimethyl ammonium bromide) Extraction buffer:

Tris Hcl	100 mM (pH-8.0)
EDTA	20 mM (pH-8.0)
NaCl	1.4M
CTAB	2% w/v
β-mercaptoethnol	0.2% (add just before extraction)

2. Phenol : chloroform: Isoamyl alcohol (25:24: 1)
3. Chilled Isopropanol
4. 70% ethanol
5. Double distilled water

Methodology

1. Take 200 mg plant sample and grid it with the help of liquid nitrogen with a pinch of PVP in mortar and pestle.
2. Take the ground sample in the autoclaved eppendorf tube and add 800 μl of CTAB extraction buffer and incubate it for 1 hour in water bath at 65^0C with intermitted shaking.
3. Then add 600 μl of phenol: chloroform: Isoamyl alcohol (25:24:1) to the eppendorf tube.
4. Mix the content of the tube by inverting tubes 2-3 times and centrifuge at 12000 rpm for 10 minutes at 40^0C.
5. Take the supernatant approx. 400-500 μl in the new autoclaved eppendorf tube and add pre chilled 600 μl Isopropanol (0.6V).
6. Mixed gently the contents of the eppendorf and centrifuge at 10000 rpm for 10 minutes at 40^0C.
7. Discard the supernatant with the help of Micropipctte and Rinse the pellet with 70% ethanol.
8. centrifuge the content of the tube at 5000 rpm for 5 minutes at 40^0C
9. Again discard the supernatant and air dry the pellet at room temperature for 15 minutes.
10. Dissolved the pellet in sterile distill water and keep it at -20°C for further use.

TO QUANTIFY DNA BY SPECTROPHOTOMETER

Requirement : Spectrophotometer, cuvett and DNA sample, sterile distill water, TE buffer.

Principle : Pure form of DNA can be measured by using a spectrophotometer. Both RNA and DNA strongly absorb the UV light. They show absorbance minimum at 230 nm and maximum 260 nm. 50 mg/ml solution of DNA (or 40 mg/ml solution of single standard RNA) gives value of 1.0 at 260 nm (A_{260}). Proteins are the main contaminants in nucleic acid extracts. Therefore, they have absorption maxima at 280 nm. A rough idea of contaminants may be have at a ratio of absorption maxima at 260 to 280 nm. Presence of fairly pure DNA and RNA indicate a ratio between 1.8 and 2.0 but the value less than 1.8 shows the presence of impurities of protein contaminants.

The following simulation illustrates the procedures for making Spectrophotometric measurements.

1. First, the intensity of light (I_0) passing through a blank is measured. The intensity is the number of photons per second. The blank is a solution that is identical to the sample solution except that the blank does not contain the solute that absorbs light. This measurement is necessary, because the cell itself scatters some of the light.
2. Second, the intensity of light (I) passing through the sample solution is measured. (In practice, instruments measure the power rather than the intensity of the light. The power is the energy per second, which is the product of the intensity (photons per second) and the energy per photon.)
3. Third, the experimental data is used to calculate two quantities: the transmittance (T) and the absorbance (A).

$$T = \frac{I}{I_o}$$

$$A = \log_{10} T$$

The transmittance is simply the fraction of light in the original beam that passes through the sample and reaches the detector. The remainder of the light, 1 - T, is the fraction of the light absorbed by the sample. Do not confuse the transmittance with the temperature, which often is given the symbol

Methodology

1. First of all switch ON the spectrophotometer and allow it to warm up to about 10 to15 minutes before use.
2. Calibrate it adjusting the wavelength at 260 nm. Switch ON the UV lamp.
3. Using distilled water (sterile) or TE buffer set the instrument at zero absorbance. This depends on there the DNA/RNA is dissolved in sterile water i.e. if DNA is dissolved in water. water must be used for adjusting the absorbancc at zero; similarly TE buffer is used for setting the same at zero when DNA is dissolved in TE buffer.

4. Note the absorbance of sample at 260 nm. If optical density (OD) is high, dilute the sample with sterile distilled water or TE buffer as per condition and read OD.
5. Calculate the concentration of double standard DNA present in sample by using the following formula.

$$\textbf{Concentration of DNA (mg/ml)} = \frac{50 \times A_{260} \times \text{dilution factor}}{1000}$$

POLYMERASE CHAIN REACTION (PCR)

Introduction

The polymerase chain reaction (PCR), now a key tool in molecular biology, provides a rapid and powerful technique for the *in vitro* amplification of DNA sequences. The PCR is now routinely used for gene cloning, analysis, and manipulation in many molecular biology laboratories. It has great potential for the study of plant pathogen interactions in facilitating both gene cloning and genome analysis.

Principle

In a typical target PCR, prior sequence information of the site is required in order to amplify the site. The sequence information is used to synthesize site specific oligonucleotide primers. At the start of the reaction, the double standard DNA is denatured by heating. The single standard template DNA is then annealed to the oligonucleotide primers that are complimentary to the sites on either side of the target locus. DNA polymerase is then used to synthesize the complimentary strand. This entire cycle results in synthesis of a replica of the target sequence. This cycle is repeated 25 to 40 times which leads to more than a million fold amplification in the copy number of the initial target sequence. PCR has been put to various uses after suitable modifications. Its amplification ranges from DNA fingerprinting using.

- Random amplification of polymorphic DNA (RAPD)
- Amplified fragment length polymorphism (AFLP)
- Simple sequence repeats (SSR)

Methodology

1. The genomic DNA isolated in the earlier experiment will be used to set up the reaction.
2. Set up a 50 μl reaction in a 0.2 ml microfuge tube

10 Taq polymerase buffer	5 μl
$MgCl_2$ (2.5 mM - mM)	2 μl
10mM dNTP	1 μl
Forward primer	1 μl
Reverse primer	1 μl
Taq polymerase (1 U/μl)	2 μl
DD Water	36 μl
Template DNA (10-50 ng)	2 μl
Total	**50 μl**

   Spin briefly (1 sec.) in a microfuge to mix.

3. PCR is performed using short cycling parameters based on Yu and Pauls (1992). Perform 25 to 35 Cycles of PCR using the following temperature profiles :

Initial denaturation	94°C, 2 min
Denaturation	94°C, 1 min.
Primer annealing	37°C, 30 sec.
Primer extension	72°C, 1 min.

   With a final extension at 72°C for 7 min. Reactions are stopped by chilling at 4°C.

4. Size fractionates the amplification products on 1.5% Agarose gel. View under UV light after ethidium bromide staining.

Agarose Gel analysis of PCR products

1. Prepare 1.5% Agarose gel in 1X TAE buffer. Allow the Agarose solution to cool to around 50°C in a water bath and add 5 μl of 10 mg/ml ethidium bromide.

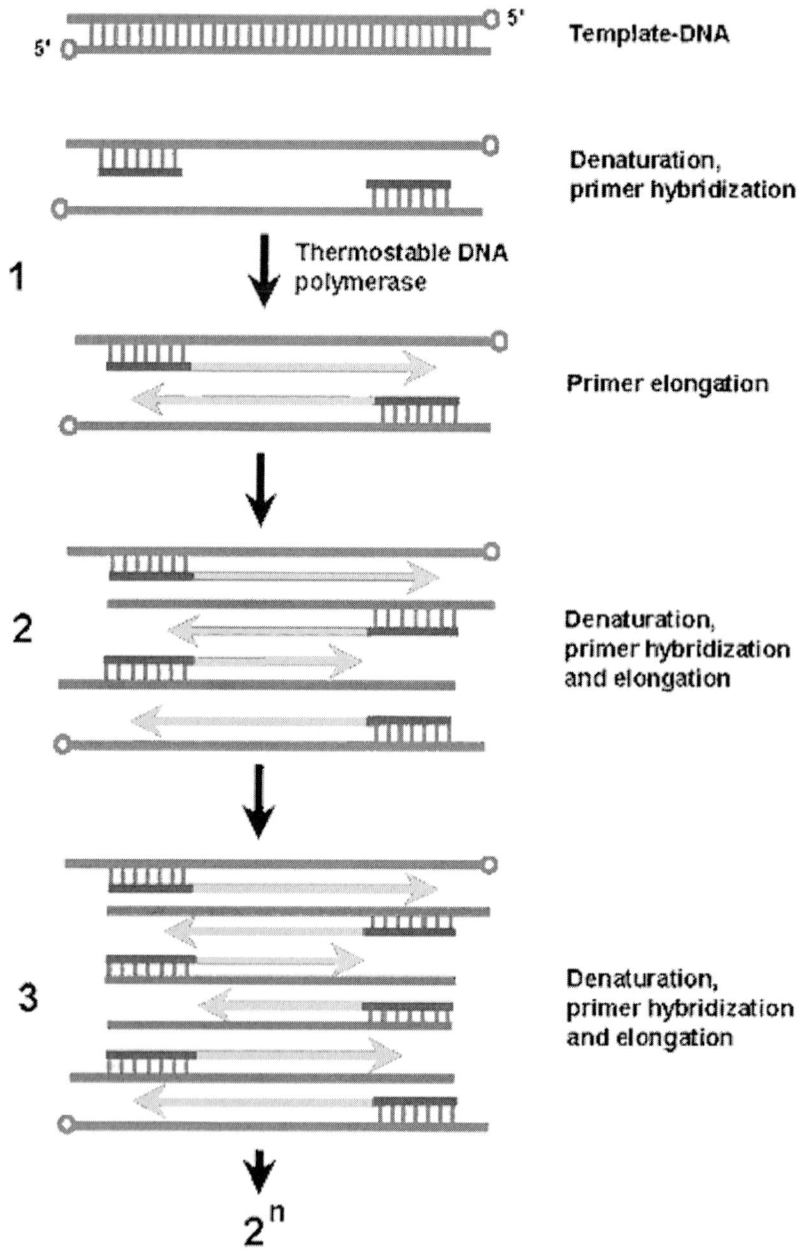

Fig. 9. Working of PCR

2. Caste a mini-gel by pouring the Agarose solution into a suitable gel mould with a suitable well-forming comb. A 5 X 7.5 cm gel can be used to separate up to 15 samples.
3. Withdraw a 5 to 10 μl sample of the PCR reaction. Add water and tracking dye. Spin the samples.
4. Store the remaining samples at 4°C or frozen.
5. Once the Agarose gel has set, remove the slot former and submerge the gel in a flat bed tank containing 1 X TAE buffer so that the gel is covered by 2 mm of buffer.
6. Load the samples and suitable molecular weight markers than do electrophoresis at 50 V for 60 min.
7. Visualize the bands on a UV transluminator and photograph if required.

Precautions

1. Put a fresh pair of gloves prior to starting the work.
2. Use fresh sterile pipette tips, PCR tube and buffers.
3. Do not use PCR reagents and solutions for other purposes.
4. Add all components of PCR reaction before adding the template DNA.
5. Keep the level of EDTA in template DNA solution as low as possible to prevent chelating of Mg^{++}.
6. Do not use excess of DNA polymerase; use dNTPs at saturating concentrations (200 μm for each dNTPs).
7. Start the PCR cycling process within 30 min. after addition of $MgCl_2$ to minimize the formation of primer dimer.

RANDOM AMPLIFIED POLYMORPHIC DNA (RAPD)

Introduction

RAPD (pronounced "rapid") stands for Random Amplification of Polymorphic DNA. It is a type of PCR reaction, but the segments of DNA that are amplified are random. The scientist performing RAPD creates several arbitrary, short primers (8 to 12 nucleotides), then proceeds with the PCR using a large template of genomic DNA, hoping that fragments will amplify. By resolving the resulting patterns, a semi-unique profile can be gleaned from a RAPD reaction.

No knowledge of the DNA sequence for the targeted gene is required, as the primers will bind somewhere in the sequence, but it is not certain exactly where. This makes the method popular for comparing the DNA of biological systems that have not had the attention of the scientific community, or in a system in which relatively few DNA sequences are compared (it is not suitable for forming a DNA databank). Because it relies on a large, intact DNA template sequence, it has some limitations in the use of degraded DNA samples. Its resolving power is much lower than targeted, species specific DNA comparison methods, such as short tandem repeats. In recent years, RAPD has been used to characterize, and trace, the phylogeny of diverse plant and animal species.

Principle

RAPD markers are decamer (10 nucleotide length) DNA fragments from PCR amplification of random segments of genomic DNA with single primer of arbitrary nucleotide sequence and which are able to differentiate between genetically distinct individuals, although not necessarily in a reproducible way. It is used for analyze the genetic diversity of an individual by using random primers. Unlike traditional PCR analysis, RAPD does not require any specific knowledge of the DNA sequence of the target organism: the identical 10-mer primers will or will not amplify a segment of DNA, depending on positions that are complementary to the primers' sequence. For example, no fragment is produced if primers annealed too far apart or 3' ends of the primers are not facing each other. Therefore, if a mutation has occurred in the template DNA at the site that was previously complementary to the primer, a PCR product will not be produced, resulting in a different pattern of amplified DNA segments on the gel.

Materials required

Random primers, 10X PCR Buffer, $MgCl_2$, dNTPs, Taq DNA polymerase. RAPD primers list are below.

Primer code	Primer Sequence (5'-3')	Annealing temperature (°C)
OPU 2	CTGAGGTCTC	32
OPU 6	ACCTTTGCGG	32
OPA 4	AATCGGGCTG	32
OPV 12	ACCCCCCACT	34
OPA 10	GTGATCGCAG	32
OPW 13	CACAGCGACA	32

Methodology

Polymerase Chain Reaction for RAPD

Preparation of master mix

Sterile distilled water	15.0 μl
10 x PCR buffer	2.5 μl
10mM dNTPs	0.5 μl
$MgCl_2$	3.0 μl
Primer (100ng/μl)	1.0 μl
Taq polymerase (1 Unit)	1.0 μl
DNA template	2.0 μl
Total PCR mix volume	**25.0 μl**

The master mix was distributed equally into microfuge tubes (23.0 μl each). 2.0 μl of template DNA was added into each microfuge according to the sample no. (All the steps were carried out in an ice box containing crushed ice). Tubes were placed in a programmed thermal cycler for 44 cycles of PCR. The programme consisted of following distinct steps :

Process	Temp.	Time duration	
Denaturation	95°C	5 min	44 cycles
Denaturation	94°C	1 min	
Annealing	30/32	1 min	
Extension	72°C	2 min	
Final extension	72°C	10 min	

After completion of the cycles, the samples were stored at 4°C till electrophoresis.

Agarose Gel Electrophoresis for Resolving PCR Amplified Product

Resolved the PCR amplified product on a 1.5% Agarose gel. Add 1.8g of Agarose to 120 ml 1X TAE buffer (prepared in sterile distilled water) and boiled the Agarose, cooled up to 50°C and add 0.2µl of ethidium bromide (10 µl /100ml stock) was added. Run the Electrophoresis was carried out at 100 Volts for 2 hour with 1X TAE as running buffer in gel electrophoresis unit . Mix (500ng) of 100 kb DNA ladder (Genei, Bangalorc) with 6X loading dye (Genei, Bangalore) to serve as molecular weight marker. Result was evaluated on a UV-transluminator.

AGAROSE GEL ELECTROPHORESIS

Introduction

Agarose gel electrophoresis is a procedure used to separate DNA fragments based on their molecular weight and is an intrinsic part of almost all routine experiment carried out in molecular biology.

This technique consists of three basic steps:

1. Preparation of Agarose gel
2. Electrophoresis of DNA fragments
3. Visualization of DNA fragments

Principle

Electrophoresis refers to movement of charged molecules under the electric field. Electrophoresis is carried out by using Agarose or polyacrylamide gel. Agarose gels are more porous (100-300nm) than that of acrylamide. Pore size depends on the concentration of Agarose. Hence, agarose is used to separate large sized macromolecules like DNA and RNA.

Agarose gel concentration depends on the size of DNA, RNA and plasmid molecules. Higher molecular weight of DNA, RNA or plasmid requires lower concentration of Agarose gel. The DNA and RNA molecules are negatively charged and moves toward anode (positive charged) when an electric field is applied. This movement depends on the size i.e. molecular weight of nucleic acid. Small sized molecules move faster than the large molecules.

Similarly, different type of plasmids is also separated by this method. A dye called ethidium bromide is used to detect the DNA. It intercalates with DNA. Therefore, location of DNA can be detected and visualized when fluoresces under UV light.

Materials required

1. Agarose
2. 50 X TAE, (Tris buffer, Glacial acetic acid, 0.5 M EDTA)
3. Distilled water

Methodology

1. Prepare 1X TAE solution from 50 x TAE stock solutions.
2. Weigh Agarose to make 0.8 % in 1 x TAE buffer.
3. Melt Agarose by boiling.
4. Before the cooling of Agarose set the caster plate and insert the comb.
5. After it has cooled to ~ 40°C and adds EtBr (ethidium bromide) to a final concentration of (0.01%).
6. Pour the Agarose onto the gel casting tray to the buffer tank.
7. After Agarose has set pour the buffer so that the gel gets submerged.
8. Load 5 μl to 10 μl samples into the wells along with a ladder (molecular weight marker).
9. Connect electrical leads to power supply and set the volt current and time.
10. Run the electrophoresis for the time period to be set.
11. Turn off power supply, disconnect leads from the input and remove.
12. Carefully put the gel on UV transluminator and with a safety shield, visualize the gel and observe the results.

PLASMID DNA ISOLATION (MINI PREP)

Introduction

Plasmids are the circular automously double stranded DNA molecules found in bacteria, cynobacteria, fungi, etc. they are classified according to the phenotypes conferred by them on the host cells e.g. antibiotic resistance, fertility factor, etc. naturally occurring plasmids also encode for genes that are required for conjugation e.g. F-plasmid. There are three forms of plasmids are used as cloning vehicle or vector to introduce foreign genes into a host organisms. There are several methods given for isolation, but the most commonly used method is called alkaline lysis.

Principle

In alkaline lysis method the cells are lysed using EDTA (those chelats metal ions) and an SDS detergent. It weakens the bacterial cell wall and also inactivates the enzyme digesting the DNA (DNase). SDS removes lipid molecules, disrupts the cell membrane and also denatures the bacterial protein.

After adding NaOH, pH of the solution increases to 11 to 12. Hence it denatures the bacterial chromosomal DNA and the plasmids DNA. When the pH is reduced after adding potassium or sodium acetate in the solution, the plasmids DNA renatures, because of its small size. But the chromosomal DNA strand and bacterial proteins form a precipitate along with SDS. You can remove the precipitate by centrifugation and adding Isopropanol and can concentrate the renatured plasmid in solution. You can remove the contaminating RNA by digesting through RNase.

Materials required

1. Luria broth media
2. Solution I (Resuspension buffer)
3. Solution II (Lysis buffer)
4. Solution III (Neutralization buffer)
5. Chilled Isopropanol
6. 70% ethanol
7. Sterile distilled water

Methodology

The most convenient method for isolation of plasmid DNA is by alkaline lysis method of Birnboim and Doly (1979).

1. Touch the bacterial colony to be screened with a sterile toothpick and inoculate into 2 ml of LB medium with appropriate antibiotic and keep in shaker incubator at 37°C, 200 rpm for 18 h. About 24 such colonies can be screened in a day.
2. Palletize the bacterial cells (next day morning) by centrifuging for 2 min in a table top centrifuge.
3. Discard the supernatant and remove the entire medium.
4. Resuspend the pellet in 100 μl of solution l. Vortex vigorously till the Pellets completely suspended.
5. Add 200 μl of solution ll and mix gently by inverting the tubes. Do not vortex.
6. Add 150 μl of chilled, solution lll and after mixing gently by inverting 2 to 3 times, keep on ice for 15 min.
7. Centrifuge at 15,000 rpm for 15 min at 4°C in a table top eppendorf centrifuge. Transfer the supernatant to a fresh tube.
8. If supernatant is found dirty, repeat the above centrifugation.
9. To the supernatant (plasmid DNA in aqueous phase), add 0.8 volume of Isopropanol to precipitate plasmid DNA and centrifuge for 15 min.
10. Discard the supernatant and to the pellet, add 200 μl of 70% ethanol. Disturb the pellet by flicking so that salt gets removed and spin for 10 min.
11. Discard the supernatant and air dry the pellet.
12. Suspend the plasmid DNA pellet in 30 μl of SDW.

RESTRICTION DIGESTION OF PLASMID WITH THE RESTRICTION ENZYME *ECO*RI

Introduction

Restriction enzyme are endonucleases which recognize and cleave the specific DNA sequence called restriction sites for example, EcoRI (isolated from Escherichia coli) that recognized and cleaves the sequence 5'-GAATTC-3' to generate cohesive or sticky ends similarly, Hind lll isolated from Haemophilus influenza and cleave the sequence 5'-AAGCTT-3' to generate cohesive or sticky ends.

Enzyme activity is represented as IU (international unit) one unit of A restriction enzyme is the amount of required to completely digest one microgram of lambda DNA (in a reaction volume of 50µl) in one hour incubation.

Principle

Restriction enzymes are a group of enzymes (called endonucleases) which recognize and cut (digest) DNA. Each restriction enzyme recognizes specific sequences and cuts at specific sites.

Example : Restriction enzyme name, recognition sequence and cutting sites are marked by " |"

Hind III	
AIAGCTT	
TTCGAIA	Sticky end or cohesive end
Bam HI	
GIGATCC	
CCTAGIG	Sticky end or cohesive end
Eco RI	
GIAATTC	
CTTAAIG	Sticky end or cohesive end
Pst I	
CTGCAIG	
GIACGTC	Sticky end or cohesive end
Eco RV	
GATIATC	
CTAITAG	Blunt end

Based on chance, one expects to find any site in a DNA sequence at a frequency of 4n, Where 4 = the number of different nucleotides in DNA (A, T, G, C) and n = the number of nucleotides in the recognition sequence. For example, for a restriction site that contains 4 base pairs, the frequency is 44 = 256. For a site composed of 6 base pairs, the frequency is 46 = 4096. A "unit" is usually defined as the amount of enzyme needed to digest 1 ug of bacterial virus lambda DNA in 1 hour in a 50 µl reaction. We generally use 5-10 units of restriction enzyme per reaction.

Materials required

1. DNA
2. 10X reaction buffer
3. Restriction enzyme
4. Sterile distill water

Methodology

Please add the following in a 1.5 ml centrifuge tube :

Sterile deionised distilled water	15.0 µl
10 X *Eco*RI enzyme buffer	2.0 µl
Restriction enzyme *Eco*RI (20 units)	1.0 µl
Plasmid DNA	2.0 µl
Total volume	**20 µl**

Maintain enzyme tube and plasmid DNA on ice. Do not leave on the Table.

1. Mix by tapping the tube with your finger.
2. Briefly centrifuge to remove bubbles (DNA will adhere to bubble surface and becomes inaccessible to the enzyme).
3. Incubate at 37°C in a water bath for 45 minutes/over night.
4. Heat inactivates the restriction enzyme by incubating the tubes at 65°C for 15 minutes.
5. Load the restriction enzyme cut and uncut plasmid on an agarose gel.

PREPARATION OF COMPETENT CELLS

Introduction

The term competent is referred to the cells which are capable to insert (receive) the vector. Transformation has been described in many species of bacteria that have the ability to take up DNA naturally. Only a small percentage of a bacterial population has this ability and these few cells are said to be competent. In E.Coli competent cells do not occur naturally, but they may so by treatment with cold $CaCl_2$ solution. After this treatment, the cells are mixed with the DNA to be introduced, given a brief heat shock and then allowed to recover. During this process, 10^7 Transformed bacteria per μg of DNA can be achieved which is enough for genetic engineering

Principle

In nature some strains of bacteria can take up DNA at stage of growth called competent stage. However many bacterial strains are not naturally competent at any stage of growth. The competent is artificially induced by treating with calcium chloride the $CaCl_2$ ions destabilizes the membrane and calcium phosphate- DNA complex is formed that adheres to the cell surface, and the DNA in taken up during heat shock and immediate chilling on ice ensures closure of the pores.

Material required

1. E.Coli strain
2. Luria broth media
3. 0.1M $MgCl_2$
4. 0.1M $CaCl_2$
5. distilled water

Methodology

1. Inoculate a single colony of E.Coli (DH5α strain into 50 ml of LB medium and allow it to grow in a shaker incubator at 37°C, 200 rpm for 18 hour (normally done in the evening).
2. Next day morning re-inoculate 0.5 ml of the overnight incubated culture into fresh 50 ml of LB broth and incubate in shaker incubator for 90 min at 37°C, 200 rpm. The log phage growth of bacteria an OD value 0.3 at 600nm is ideal for transformation.
3. From the step onwards carry out the entire step in a laminar air flow cabinet providing aseptic conditions. After obtaining the growth, transfer the culture to 50 ml polypropylene tubes under aseptic condition and keep on ice for 10 minutes. Palletize the bacterial cells by

centrifuging at 5000 rpm for 10 minutes at 4°C. Discard the medium and suspend the cells gently in 10ml of ice cold solution of 0.1 M $MgCl_2$.

4. Keep the tubes containing cells on ice for 1 hour (1 $CaCl_2$ treatment).
5. Recover the cells by centrifugation at 5000 rpm at 40C for 10 minutes. Resuspend the cells by gently pipetting up and down in 0.5ml of chilled 0.1 M $CaCl_{2.}$
6. Keep the cells on ice for 1 hour (2nd $CaCl_2$ treatment).

Cells are now competent. Either they can be used immediately or stored. For storing, 200 μl of cells, 50% glycerol and used plunge it in liquid nitrogen to freeze them. Store at -70°C in eppendorf tubes. Whenever required, the tubes are kept on ice allowed to thaw gradually and use. Avoid frequently freezing and thawing.

MOLECULAR CLONING OF GENE

Introduction

"Gene cloning is molecular method of isolation and multiplication of an individual gene sequence by its insertion into a bacterial cell, which can multiply the gene as it multiple itself". Gene cloning is a straight forward procedure. A fragment of DNA, containing the gene to be cloned, is inserted into a circular DNA molecular called a vector to procedure a chimera or recombinant DNA molecules. The vector acts as a vehicle that transports the gene into host cells, which is usually a bacterium, although other type s of living cells can be used. Within the host cell the vector multiples producing numerous identical copies not only of itself but also of the gene that it carries. When the host cell divides copies of the recombinants DNA molecules are passed to the progeny and further vector replication takes places. After a large number of cells divisions, a colony or clone of identical host cells are produced. Each cell in the clone contains one or more copies of the recombinants DNA molecule. The gene carried by the recombinant molecule is now said to be cloned

Principle

To clone any insert, we require,

1. Purified DNA of the organism (fungal/ bacterial/ viral/ amplified gene).
2. A Cloning vector- the vector is a phage or plasmid that is used to carry the inserted foreign DNA for the purpose of producing more copies. The plasmids presently used are genetically engineered derivatives of naturally occurring plasmids which are double-stranded and autonomously replicating.

   In order to facilitate cloning, they have the following features:

(a) They have bacterial origin of replication.

(b) They bear antibiotic resistance genes serving as markers.

(c) In the case of M13 coli phage derived vectors, they have sequences coding for α-chain of *Lac Z* gene.

(d) All the plasmids have a short stretch (~ 50 nt.) of sequences that are recognized by specific restriction enzymes. These sites do not occur anywhere else in the vector and so called as unique sites for specific enzymes. It is these sites which are used for cloning. The 50 nt stretch having the unique sites for selected restriction enzymes are called multiple cloning site or polycloning site. Any DNA that is to be cloned is restricted by an enzyme, the vector is also restricted by the same enzyme and the foreign DNA is inserted between the two ends of the vector at the restriction enzyme site multiple cloning sites. The gaps between the two DNA fragments are sealed by the enzyme ligase.

(e) The new DNA molecule, hybrid DNA is called recombinant DNA molecule which can be introduced into Escherichia coli. The bacterial cells having recombinant molecule are selected as below :

(f) Using the antibiotic marker gene : When the transformed bacteria are plated on media containing selected antibiotics, only the cells having recombinant molecules will survive.

(g) It is also selected by histochemical method – essentially called as α-complementation. Many of the M13 derived vectors have a short segment of E. coli DNA which has the regulatory sequences and the coding sequence for the first 146 amino acids of β-*glycosidase* gene (lac Z). The polycloning site lies embedded in this coding region, but does not disrupt the reading frame and results in the harmless interpolation of a few AAs in the N-terminal fragment of coding for the C-terminus of the gene. These two peptides coded by the vector and host are not active enzyme. This is called as β-complementation. The Lac^+ bacteria form blue colonies when IPTG (Isopropyl thio galactoside) and X-gal (5-bromo-4-Chloro-3-indolyl-β-D-galactoside) are added. But insertion of the foreign DNA into the multiple cloning sites present in the coding region disrupts the frame of the protein and produces pale blue or white colonies.

Methodology

Preparation of genomic DNA and vector DNA for cloning : Genomic DNA prepared earlier will be used for cloning. The vector pUC 18 obtained from commercial source will be used.

I. Restriction of genomic DNA

Prepare the reaction mix as follows. Add the components one after another in the order given below. Always write down the volume of the components on your lab notebook and then start. Have an ice bucket with ice. Immerse the eppendorf tube in which the reaction to be conducted on ice.

DNA	10 µl (~ 10 µg)
SDW	78 µl
10X Reaction buffer	10 µl
Restriction enzyme(*Hind III/Bam HI/Eco RI/Pst I*)	2 µl
Total	**100 µl**

1. Follow the enzymes manufacturer's instructions for the temperature and buffer requirements.
2. Always take the enzymes from -20°C only for a brief time. As soon as possible keep it on ice while using.
3. Mix the ingredients by flicking the eppendorf tube.
4. Briefly centrifuge the samples to bring the liquids to the bottom of the tube.
5. Incubate the mixture at 37°C for 2 to 3 hrs.
6. Prepare 1% Agarose gel with 0.01% ethidium bromide.

7. At the end of incubation, remove 5 µl of the restricted genomic DNA sample and analyze the extent of digestion by gel electrophoresis. A uniform smear throughout the lane is indicative of complete digestion and the generation of a lot of fragments.

If the digestion is complete, heat inactivate the enzymes (*Eco RI/Pst I*) by keeping water bath at 65°C for 30 min. If the enzymes is not heat inactivatable (*Hind III/Bam HI*), remove the enzymes by deproteinisation using phenol: Chloroform: Isoamyl alcohol (25:24:1) as below.

1. To thc sample, add the required amount of sterile distilled water (SDW) to make unto 200 µl volumes.
2. Add equal amount of Phenol: Chloroform: Isoamyl alcohol, i.e. 200 µl, gently mix and spin for 10 min, at 15,000 rpm.
3. Take the upper aqueous phase and add equal amount of C: I (24: 1) and spin for 10 min.
4. Take the upper aqueous phase and add 1/10 volume of 3M sodium acetate pH 4.8 and 0.8 volume of Isopropanol or 2.5 volume of ethanol.
5. Mix well and keep the sample at -20°C for 1 hr.
6. Spin at 15,000 rpm for 15 min. in refrigerated microfuge.
7. Wash the pellet with 70% ethanol by adding 500 µl of it and spinning for 15 min.
8. Discard the alcohol and air dry the pellet.
9. Dissolve the pellet in 25 µl of SDW.
10. Prepare 1% Agarose gel with 0.01% ethidium bromide.
11. Load 5 µl of the restricted foreign DNA on 1% agarose gel and run at 75V and view under UV to assess the quality of extracted DNA. 5 µl aliquot shows good fluorescence.

II. Restriction of plasmid pUC 18 DNA

Plasmid DNA	4 µl (~ 2 µg)
SDW	15 µl
10X restriction buffer	2 µl
Enzyme	1 µl (5-10 U)
Total	**20 µl**

1. Mix the ingredients by tapping the eppendorf.
2. Briefly centrifuge the samples.
3. Incubate at 37°C for 2 hrs.
4. Meanwhile, prepare 1% Agarose gel with 0.01% EtBr.
5. At the end of incubation, treat the vector with calf intestinal alkaline phosphatase (CEAP) [to remove 5 phosphate group and to prevent recircularisation of the vector during ligation as the ends generated by the digestion are compatible] i.e. add 10X CIAP buffer along with 3-4 units of the enzyme, CIAP.
6. Mix well and incubate the sample at 37°C for 1 h.
7. Remove the enzymes by extraction with phenol chloroform: Isoamyl alcohol (as given earlier).
8. Dissolve the pellet in 20 µl of SDW (100 ng/ µl of DNA).
9. To check the concentration and quality of the linearised vector DNA, load 1 µl of vector DNA on 1% Agarose gel and run at 75 V.
10. View Under UV and assess the quality of DNA, the plasmid DNA should appear as single band. Compare the size with molecule weight marker. In pUC 18, the linear band will be ~ 2.7 kb.

III. Ligation of fungal DNA into the vector pUC 18

Prepare the ligation mix as follows:

Linearised, dephosphorylatedVector DNA	2 µl (~ 150-200 ng)
Restricted fungal DNA -	20 µl (~ 500-1 µg)
SDW	1 µl
10X Ligase buffer	3 µl
10 mM ATP	3 µl
T_4 DNA ligase	1 µl (1 wiess unit)
Total	**30 µl**

1. Mix the contents gently and briefly spin down.
2. Incubate at 15°C for 18 h or overnight.

Any DNA fragment can be blunt ended with T4 DNA polymerase and legated with Sma I cut vector to give rise to recombinant molecule. Following is the procedure:

1. Cut the specific DNA fragment from the gel.
2. Purify the DNA from the gel using gel purification kit.
3. Check the concentration of the DNA in gel electrophoresis to ~ 250-300 ng of DNA add the following :

Gel extracted DNA	10 μl (~ 300 ng of DNA)
SDW	5 μl
10X reaction buffer	2 μl
dNTP	2 μl
T4 DNA polymerase	1 μl
Total	**20 μl**

4. Incubate at 11°C for 20 min.
5. Heat inactivates T4 DNA polymerase by keeping the reaction mix at 65°C for 30 min.
6. To above blunt ended DNA add the following for ligation.

Blunt ended DNA	20 μl
Sma I/CIAP treated pUC 18	3 μl
SDW	0 μl
10X ligase buffer	3 μl
10 mM ATP	3 μl
Ligase	1 μl
Total	**30 μl**

7. Incubate at 15°C overnight.
8. Use this also for transformation.

Transformation

Transformation has been described in many species of bacteria that have the ability to take up DNA naturally. Only a small percentage of a bacterial population has this ability and these few cells are said to be competent. In E. coli, competent cells do not occur naturally, but they may be so by treatment with cold $CaCl_2$ solutions. After this treatment, the cells are mixed with the DNA to be introduced, given a brief heat shock and then allowed to recover. During this process, 10^7 transformed bacteria per μg of DNA can be achieved which is enough for genetic engineering.

I. Preparation of competent cells by $CaCl_2$ method (Mandel & Higa, 1970)

1. Inoculate a single colony of E. coli (DH5α strain) into 50 ml of LB medium and allow it to grow in a shaker incubator at 37°C, 200 rpm for 18 h (normally done in the evening).
2. Next day morning, re-inoculate 0.5 ml of the overnight incubated culture into fresh 50 ml of LB broth and incubate in shaker incubator for 90 min at 37°C, 200 rpm.
3. The log phase growth of bacteria at OD value 0.3 at 600 nm is ideal for transformation.
4. From this step onwards, carry out all the steps in a laminar air flow cabinet providing aseptic conditions. After obtaining the growth, transfer the culture to 50 ml polypropylene tubes under aseptic conditions and keep on ice for 10 min.
5. Palletize the bacterial cells by centrifuging at 5000 rpm for 10 min at 4°C.
6. Discard the medium and suspend the cells gently in 10 ml of ice cold solution of 0.1 M $MgCL_2$.
7. Keep the tunes containing cells on ice for 1 h ($MgCl_2$ treatment).
8. Recover the cells by centrifugation at 5000 rpm at 4°C for 10 min.
9. Resuspend the cells by gently pipetting up and down in 0.5 ml of chilled 0.1 M $CaCl_2$.
10. Keep the cells on ice for 1 hr (ll $CaCl_2$ treatment).

Cells are now competent. Either they can be used immediately or stored. For storing to 200 μl of cells, add 10% glycerol and plunge it in liquid nitrogen to freeze them. Store them at -70°C in eppendorf tubes. Whenever required, the tubes are kept on ice allowed to thaw gradually and used. Avoid frequent freezing and thawing.

II. Procedure for transforming bacterial cells

1. Add 200 μl of competent cells to the ligation mix, mix gently in the eppendorf and incubate on ice for 1h.
2. Suspend the eppendorf tube with cells and ligation mix in a water bath at 42°C for 2 min (heat shock).
3. Place the tubes on ice and take it to aseptic condition and add 1 ml of LB medium and keep in shaker incubator at 37°C, 200 rpm for 1h.
4. Harvest the bacterial cells by centrifugation in a microfuge for 1 min.
5. Take only 200 μl of the cell suspension and spread it on the appropriate, selection medium i.e. LA plates containing the required antibiotic (50 μg of ampicillin per ml), 20 μl of 0.1 M IPTG and 200 μl of 2% X-gal for 100 ml LA medium, with the help of a spreader.
6. Seal the plates and incubate overnight in inverted position at 37°C.

Following controls will have to be maintained. They are processed the same way as DNA samples used for cloning.

(a) Unrestricted plasmid vector (1 ng DNA). They are supercoiled intact DNA, they will enter the cells efficiently. As the lac Z operon is not disturbed, α complementation will take place, so it will give rise to blue colony.

(b) Restricted plasmid vector, dephosphorylated, ligase added as both the ends have OH ends, they cannot join. So no colony should be seen.

(c) Plain cells as such spread. No growth should be seen as antibiotic resistance is not there in the bacterial chromosomal DNA.

(d) Restricted plasmid vector, not dephosphorylated but ligase added. It has 5' phosphate and 3' OH end, it should ligate and give rise to a lot of blue colonies, if ligase is good. This is done to test the ligase used.

SELECTION OF RECOMBINANT CLONES BY RESTRICTION ENZYME AND COLONY PCR

Introduction

Selections of recombinant clone on the basis of the screening of blue /white colony after it we will be confirm the transformation by the restriction digestion or colony PCR. Restriction enzyme is endonucleases which recognize and cleave the specific DNA sequence called restriction sites for example, *EcoRI* Even though blue/white screening can be used to determine if inserts are present, colony PCR can be used to determine insert size and/or orientation in the vector. Alternately, the presence of an insert and its size can be determined by growing each colony in liquid, the plasmid purified by boiling or alkaline preparation protocol, digestion of the plasmid with restriction enzyme(s) that excises the insert, followed by separation by Agarose gel electrophoresis.

Principle

X-Gal (5-bromo -4-chloro-3-indolyl-β-D- galacto-pyranoside)is an inert chromogenic substrate for β-galactisidase which hydrolyzes X-gal into colorless galactose and 4-chloro-3-bromo-indigo, forming an intense blue precipitate. Induction of the lacZ gene with IPTG leads to the hydrolysis of X-gal solution, ready to use, is stable, 0.22 μm membrane filtered especially formulated for blue/white colony screenings assays. IPTG (isopropyl-β-D-thio-galacto-pyranoside) is a synthetic analog of lactose, it inactivate the lac repressor and induces synthetic of β-galactosidase an enzyme that promotes lactose utilization. IPTG is used to induce expression of cloned genes under control of the lac operon, it is used in conjunction with X-gal to determine the lac phenotype in blue/white colony screening.

Materials

1. Luria agar
2. Luria broth
3. Ampicillin
4. X-gal
5. IPTG

Screening & Analysis of Recombinant Clone

1. Inoculate 10-20 white colonies into 5 ml LB medium containing 100 mg/ml of ampicillin and 40 μg/ml X-gal and 40 μl of 100 mM of IPTG solution, ready to. Incubate overnight on shaker at 37°C.
2. Next day we observe the blue white colony in the plate.
3. Do not screen cultures that develop blue color. These are false positive.

4. Select only white colony for the plasmid isolation.
5. Analyze by restriction digestion / colony PCR
6. Analysis 4-6 colony for the presence and orientation of the DNA insert using one of the following methods.

Restriction digestion

1. Using the appropriate enzyme for the restriction digestion and see the methodology on page no.
2. Agarose analysis of restriction digests.

Colony PCR

This protocol is designed to quickly screen for plasmid inserts directly from E. coli colonies. The plasmid should be high copy number such pUC 18, pUC 19, etc. Even though blue/white screening can be used to determine if inserts are present, this technique can be used to determine insert size and/or orientation in the vector. Alternately, the presence of an insert and its size can be determined by growing each colony in liquid, the plasmid purified by boiling or alkaline preparation protocol, digestion of the plasmid with restriction enzyme(s) that excises the insert, followed by separation by Agarose gel electrophoresis.

Typical colony PCR reaction

Mix together the following on ice; always adding enzyme last. For multiple samples, make a large master mix and aliquot 50 μl in each PCR tube (also on ice).

Sterile distilled water	38 μl
10XPCR buffer	5 μl
25 mM $MgCl_2$	3 μl
10 mM dNTPs	1 μl
20 μM forward primer	1 μl
20 μl reverse primer	1 μl
Taq polymerase	0.2-1 μl
Total volume	**50 μl**

To each cold PCR tube containing the PCR reaction, add a small amount of colony. To do this, use a fine yellow pipette tip attached to a pipette (set at 30 μlto avoid addition of air into the PCR reaction) and pipette up and down to mix. The amount of cells should be small, just a touch will do, the small amount required to fill the end of the opening is sufficient. Sufficient mixing will result in complete cell lysis and high yields.

PCR conditions	Time and Temperature	What happens?
1 cycle	5 min at 95°C	initial cell breakage and DNA denaturation
	1 min at 95°C	DNA denatures into single strands
30-40 cycles	1.5 min at 54°C	primers anneal to ssDNA template (temperature depends on primers)
	1 min at 72°C	primes are extended from 3′ end by Taq(1 min/kb)
1 cycle	5 min at 72°C	final extension to make sure all products are full length (72°C is optimal for Taq polymerase)

6 Tissue Culture Techniques

TISSUE CULTURE METHODS

Types of Cells Grown in Culture

Tissue culture refers to both organ culture and cell culture. Cell cultures are derived from either primary tissue explants or cell suspensions. Primary cell cultures typically will have a finite life span in culture, whereas continuous cell lines are, by definition, abnormal and are often transformed cell lines.

Work Area and Equipment

(a) **Laminar Flow Hoods :** The laminar flow hoods are of 2 types, vertical and horizontal. The vertical hood is best for working with hazardous organisms, since the aerosols that are generated in the hood are filtered out before they are released into the surrounding environment. Horizontal hoods are designed so that the air flows directly at the operator, hence, they are not useful for working with hazardous organisms, but are the best protection for your cultures. Both types of hoods have continuous displacement of air that passes through a HEPA (high efficiency particle) filter that removes particulates from the air. The hoods are equipped with a shortwave UV light that can be turned on for a few minutes to sterilize the surfaces of the hood, but be aware that only exposed surfaces will be accessible to the UV light. Do not put your hands or face near the hood when the UV light is on, as the shortwave light can cause skin and eye damage. The hoods should be turned on about 10 to 20 minutes before being used. Wipe down all surfaces with ethanol before and after each use.

(b) **Microscopes :** Inverted phase contrast microscopes are used for visualizing the cells. Microscopes should be kept covered and the lights turned down when not in use. Before using the microscope or whenever an objective is changed, check that the phase rings are aligned.

(c) **CO_2 Incubator :** The cells are grown in an atmosphere of 5% to 10% CO_2, because the medium used is buffered with sodium bicarbonate/carbonic acid and the pH must be strictly maintained. Culture flasks should have loosened caps to allow for sufficient gas exchange. The humidity must also be maintained for those cells growing in tissue culture dishes, so a pan of water is kept filled at all times.

(d) **Preservation :** Cells are stored in liquid nitrogen.

(e) **Vessels :** Anchorage-dependent cells require a nontoxic, biologically inert, and optically transparent surface that will allow cells to attach and allow movement for growth. The most convenient vessels are specially-treated polystyrene plastic that are supplied sterile and are disposable. These include petri dishes, multiwell plates, microtiter plates, roller bottles, and screwcap flasks.

Preservation and Storage

Liquid N_2 is used to preserve tissue culture cells, either in the liquid phase or in the vapor phase. Freezing can be lethal to cells due to the effects of damage by ice crystals, alterations in the concentration of electrolytes, dehydration, and changes in pH. To minimize the effects of freezing, several precautions are taken. First, a cryoprotective agent that lowers the freezing point, such as glycerol or DMSO, is added. The freezing medium is typically 90% serum, 10% DMSO. In addition, it is best to use healthy cells that are growing in log phase and to replace the medium 24 hours before freezing. Also, the cells are slowly cooled from room temperature to –80°C to allow the water to move out of the cells before it freezes. The optimal rate of cooling is 1°C–3°C per minute. Some labs have fancy freezing chambers to regulate the freezing at the optimal rate by periodically pulsing in liquid nitrogen. The tubes filled with 200 ml of isopropanol at room temperature and the freezing vials containing the cells are placed in the container. The container is placed in the –80°C freezer. The effect of the isopropanol is to allow the tubes to come to the temperature of the freezer slowly, at about 1°C per minute. Once the container has reached –80°C the vials are removed and immediately placed in the liquid nitrogen storage tank. Cells are stored at liquid nitrogen temperatures because the growth of ice crystals is retarded below –130°C.

Maintenance

Cultures should be examined daily, observing the morphology, the color of the medium, and the density of the cells. A tissue culture log should be maintained. The log should contain the name of the cell line, the medium components, and any alterations to the standard medium, the dates on which the cells were split and/or fed, a calculation of the doubling time of the culture (this should be done at least once during the semester), and any observations relative to the morphology, etc.

(a) **Growth Pattern :** Cells will initially go through a quiescent or lag phase that depends on the cell type, the seeding density, the media components, and previous handling. The cells will then go into exponential growth where they have the highest metabolic activity. The cells will then enter into stationary phase where the number of cells is constant. This is characteristic of a confluent population (where all growth surfaces are covered).

(b) **Harvesting :** Cells are harvested when the cells have reached a population density that suppresses growth. Ideally, cells are harvested when they are in a semiconfluent state and are still in log phase. Cells that are not passaged and are allowed to grow to a confluent state can sometime lag for a long period of time, and some may never recover. It is also essential to keep your cells as happy as possible to maximize the efficiency of transformation. Most cells are passaged (or at least fed) 3 times a week.

1. **Suspension cultures :** Suspension cultures are fed by dilution into a fresh medium.

2. **Adherent cultures :** Adherent cultures that do not need to be divided can simply be fed by removing the old medium and replacing it with fresh medium. When the cells become semiconfluent, several methods are used to remove the cells from the growing surface so that they can be diluted:

Mechanical A rubber spatula can be used to physically remove the cells from the growth surface. This method is quick and easy, but is also disruptive to the cells and may result in significant cell death. This method is best when harvesting many different samples of cells for preparing extracts, i.e., when viability is not important.

Proteolytic enzymes Trypsin, collagenase, or pronase, usually in combination with EDTA, causes cells to detach from the growth surface. This method is fast and reliable but can damage the cell surface by digesting exposed cell surface proteins. The proteolysis reaction can be quickly terminated by the addition of complete medium containing serum. EDTA alone can also be used to detach cells and seems to be gentler on the cells than trypsin.

Safety Considerations

Assume all cultures are hazardous since they may harbor latent viruses or other organisms that are uncharacterized. The following safety precautions should also be observed:

1. Use pipette aids to prevent ingestion and keep aerosols down to a minimum.
2. No eating, drinking, or smoking.
3. Wash hands after handling cultures and before leaving the lab.
4. Decontaminate work surfaces with disinfectant (before and after).
5. Autoclave all waste.
6. Use biological safety cabinet (laminar flow hood) when working with hazardous organisms. The cabinet protects worker by preventing airborne cells and viruses released during experimental activity from escaping the cabinet; there is an air barrier at the front opening and exhaust air is filtered with a HEPA filter.

7. Make sure the cabinet is not overloaded and leave exhaust grills in the front and the back clear (helps to maintain a uniform airflow).
8. Use aseptic technique.
9. Dispose of all liquid waste after each experiment and treat with bleach.

TISSUE CULTURE METHODS

Each student should maintain his or her own cells throughout the course of the experiment. These cells should be monitored daily for morphology and growth characteristics, fed every 2 to 3 days, and subcultured when necessary. A minimum of two 25-cm^2 flasks should be carried for each cell line; these cells should be expanded as necessary for the transfection experiments. Each time the cells are subcultured, a viable cell count should be done, the subculture dilutions should be noted, and after several passages, a doubling time determined. As soon as you have enough cells, several vials should be frozen away and stored in liquid N_2. One vial from each freeze down should be thawed 1 to 2 weeks after freezing to check for viability. These frozen stocks will prove to be vital if any of your cultures become contaminated.

Procedures

1. **Media preparation :** Each student will be responsible for maintaining his or her own stock of cell culture media; the particular type of media, the sera type, and concentration, and other supplements will depend on the cell line. Do not share media with your partner (or anyone else), because if a culture or a bottle of media gets contaminated, you have no back-up. Most of the media components will be purchased prepared and sterile. In general, all you need to do is sterilely combine several sterile solutions. To test for sterility after adding all components, pipette several mL from each media bottle into a small sterile Petri dish or culture tube and incubate at 37°C for several days. Use only media that has been sterility-tested. For this reason, you must anticipate your culture needs in advance so you can prepare the reagents necessary. But, please, try not to waste media. Anticipate your needs but don't make more than you need. Tissue culture reagents are very expensive; for example, bovine fetal calf

serum cost ~$200 per 500 ml. Some cell culture additives will be provided in a powdered form. These should be reconstituted to the appropriate concentration with double-distilled water (or medium, as appropriate) and filtered (in a sterile hood) through a 0 to 22 mm filter.

All media preparation and other cell culture work must be performed in a laminar flow hood. Before beginning your work, turn on the blower for several minutes; wipe down all surfaces with 70% ethanol, and ethanol washes your clean hands. Use only sterile pipettes, disposable test tubes and autoclaved pipette tips for cell culture. All culture vessels, test tubes, pipette tip boxes, stocks of sterile eppendorfs, etc., should be opened only in the laminar flow hood. If something is opened elsewhere in the lab by accident, you can probably assume it's contaminated. If something does become contaminated, immediately discard the contaminated materials into the biohazard container and notify the instructor.

2. **Growth and morphology :** Visually inspect cells frequently. Cell culture is sometimes more an art than a science. Get to know what makes your cells happy. Frequent feeding is important for maintaining the pH balance of the medium and for eliminating waste products. Cells do not typically like to be too confluent, so they should be subcultured when they are in a semiconfluent state. In general, mammalian cells should be handled gently. They should not be vortexed, vigorously pipetted or centrifuged at greater than 1500 g.
3. **Cell feeding :** Use prewarmed media and have cells out of the incubator for as little time as possible. Use 10 to15 ml for T-25s, 25–35 ml for T-75s, and 50–60 mL for T-150s.
   a. **Suspension cultures :** Feeding and subculturing suspension cultures are done simultaneously. About every 2 to 3 days, dilute the cells into fresh media. The dilution you use will depend on the density of the cells and how quickly they divide, which only you can determine. Typically 1: 4 to 1:20 dilutions are appropriate for most cell lines.
   b. **Adherent cells :** About every 2 to 3 days, pour off old media from culture flasks and replace with fresh media. Subculture cells as described below before confluency is reached.

4. **Subculturing adherent cells** When adherent cells become semiconfluent, subculture using 2 mm EDTA or trypsin/EDTA.

Trypsin-EDTA :

a. Remove medium from culture dish and wash cells in a balanced salt solution without Ca^{++} or Mg^{++}. Remove the wash solution.

b. Add enough trypsin-EDTA solution to cover the bottom of the culture vessel and then pour off the excess.

c. Place culture in the 37°C incubator for 2 minutes.

d. Monitor cells under microscope. Cells are beginning to detach when they appear rounded.

e. As soon as cells are in suspension, immediately add the culture medium containing serum. Wash cells once with serum-containing medium and dilute as appropriate (generally 4- to 20-fold).

EDTA alone :

a. Prepare a 2 mm EDTA solution in a balanced salt solution (i.e., PBS without Ca^{++} or Mg^{++}).

b. Remove the medium from the culture vessel by aspiration and wash the monolayer to remove all traces of serum. Remove salt solution by aspiration.

c. Dispense enough EDTA solution into culture vessels to completely cover the monolayer of cells.

d. The coated cells are allowed to incubate until cells detach from the surface. Progress can be checked by examination with an inverted microscope. Cells can be gently nudged by banging the side of the flask against the palm of the hand.

e. Dilute cells with fresh medium and transfer to a sterile centrifuge tube.

f. Spin cells down, remove supernatant, and resuspend in culture medium (or freezing medium if cells are to be frozen). Dilute as appropriate into culture flasks.

5. **Thawing frozen cells**

a. Remove cells from frozen storage and quickly thaw in a 37°C waterbath by gently agitating vial.

b. As soon as the ice crystals melt, pipette gently into a culture flask containing prewarmed growth medium.

c. Log out cells in the “Liquid Nitrogen Freezer Log” Book.

6. **Freezing cells**
   a. Harvests cells as usual and wash once with complete medium.
   b. Resuspend cells in complete medium and determine cell count/ viability.
   c. Centrifuge and resuspend in ice-cold freezing medium: 90% calf serum/ 10% DMSO, at 106–107 cells/ml. Keep cells on ice.
   d. Transfer 1-mL aliquots to freezer vials on ice.
   e. Place in a Mr. Frosty container that is at room temperature and has sufficient isopropanol.
   f. Place the Mr. Frosty in the –70°C freezer overnight. Note: Cells should be exposed to freezing medium for as little time as possible prior to freezing.
   g. The next day, transfer to liquid nitrogen (DON'T FORGET) and log in the "Liquid Nitrogen Freezer Log" Book.
7. **Viable cell counts** Use a hemocytometer to determine total cell counts and viable cell numbers. Trypan blue is one of several stains recommended for use in dye exclusion procedures for viable cell counting. This method is based on the principle that live cells do not take up certain dyes, whereas dead cells do.
   a. Prepare a cell suspension, either directly from a cell culture or from a concentrated or diluted suspension (depending on the cell density) and combine 20 ml of cells with 20 ml of trypan blue suspension (0.4%). Mix thoroughly and allow to stand for 5–15 minutes.
   b. With the cover slip in place, transfer a small amount of trypan blue-cell suspension to both chambers of the hemocytometer by carefully touching the edge of the cover slip with the pipette tip and allowing each chamber to fill by capillary action. Do not overfill or under-fill the chambers.
   c. Starting with 1 chamber of the hemocytometer, count all the cells in the 1-mm center square and each of the four 1-mm corner squares. Keep a separate count of viable and nonviable cells.
   d. If there are too many or too few cells to count, repeat the procedure, either concentrating or diluting the original suspension as appropriate.

e. The circle indicates the approximate area covered at 100X microscope magnification (10X ocular and 10X objective). Include cells on top and left, touching the middle line. Do not count cells touching the middle line at the bottom and right. Count 4 corner squares and the middle square in both chambers and calculate the average.

f. Each large square of the hemocytometer, with cover slip in place, represents a total volume of 0.1 mm^3 or 10 to 4 cm^3. Since 1 cm^3 is equivalent to approximately 1 ml, the total number of cells per ml will be determined using the following calculations:

Cells/ml = average cell count per square × dilution factor × 10^4; Total cells = cells/ml × the original volume of fluid from which the cell sample was removed; % Cell viability = total viable cells (unstained)/total cells × 100.

Plant Growth Regulators Commonly Used in Plant Tissue Culture

Class	Name	Abbreviation	Comments
Auxin	Indole-3-acetic acid	IAA	Use for callus induction at 10 30 mM. Lowering to 1 to 10 mM can stimulate organogenesis. It is inactivated by light and readily oxidized by plant cells. The synthetic auxins below have largely superceded IAA for tissue culture studies.
	Indole-3-butyric acid	IBA	Use for rooting shoots regenerated via organogenesis. Either maintain at a low concentration (1 to 50 mM) throughout the rooting process, or expose to a high concentration (100 to 250 mM) for 2 to 10 days and then transfer to a hormone-free medium. It can also be used as a dip for in vitro or ex vitro rooting of shoots.

[Table Contd...

Contd. Table]

Class	Name	Abbreviation	Comments
	2,4-Dichlorophenoxy acetic acid	2,4-D	Most commonly used synthetic auxin for inducing callus and maintaining callus and suspension cells in dedifferentiated states. Usually used as the sole auxin source (1 to 50 mM), or in combination with NAA.
	p-Chlorophenoxyacetic acid 1-Naphthalene acetic acid		pCPA NAA Similar to 2, 4-D, but less commonly used Synthetic analogue of IAA. Commonly acetic acid used either as the sole auxin source (2 to 20 mM for callus induction and growth of callus and suspension cultures, 0.2 to 2 mM for root induction), or in combination with 2,4-D.
Cytokinin	6-Furfurylamino- purine (kinetin)	K	Often included in culture media for callus induction, growth of callus and cell suspensions, and induction of morpho-genesis (1 to 20 mM). Higher concentrations (20 to 50 mM) can be used to induce the rapid multiplication of shoots, axillary/adventitious buds or meristem.
	6-Benzylaminopurine	BAP, BA	Included in culture media for callus induction, growth of callus and cell suspensions (0.5 to 5.0 mM), and for induction of morphogenesis (1 to 10 mM). More commonly used than kinetin for inducing rapid multiplication of shoots, buds, or meristems (5 to 50 mM).

[Table Contd...

Contd. Table]

Class	Name	Abbreviation	Comments
	N-Isopenteny-laminopurine	2iP	Less commonly used than K or BAP for callus induction and growth (2–10 mM), induction of morphogenesis (10 to 15 mM), or multiplication of shoots, buds, or meristems (30 to 50 mM).
	Zeatin	Zea	Seldom used in callus or suspension media. Can be used for induction of morphogenesis (0.05 to 10 mM). Zea is thermolabile and must not be autoclaved.
Gibberellin	Gibberellin A3	GA3	Seldom used in callus or suspension medium (one exception being potato). Can promote shoot growth when added to shoot induction medium at 0.03 to 14 mM. Also used to en hance development in embryo/ ovule cultures (0.3 to 48 mM). GA3 is thermolabile and must not be autoclaved.
Abscisic	Abscisic acid	ABA	Used at concentrations of 0.4 to 10mM to prevent precocious germination, and promote normal development of somatic embryos.

PLANT TISSUE CULTURE

Introduction

Plant tissue culture involves the growth of plant cells, tissues or segments for purposes such as genera ting or cloning large amounts of new cells, tissues or plants; to study development; to provide mechanisms of genetic engineering; or to produce valuable chemicals found in plant cells. Cloning is not an alien concept to those people familiar with plants. It is a process of generating exact and multiple copies of a parent plant, Cloning can be done by making cuttings, grafts, or even naturally. Plant tissue culture uses much smaller pieces of tissue in this cloning process. Of course, there are so many possibilities in this field that plant tissue culturists can do more than simply clone plants. Plant tissue culturists perform a number of different types of experiments to place new plants into culture or to develop new methods of working with plant cultures.

The plant tissue culturist should have a good idea at the outset as to the purpose of the experiments. There are a number of different types of *in vitro* tissue culture approaches. For tissue culture, the term *in vitro* refers to growing picces of plants separate from the entire organism. We will explore some of these in our laboratory. Before we begin our experiments, it is helpful to describe some of the more important types of tissue culture.

MICROPROPAGATION

Micropropagation uses small pieces of tissues such as axillary buds, tubers or rhizomes for rapid cloning or generation of new plants. Think of a bud on a plant. The bud contains the meristematic area. Under *in vivo* (on the plant) conditions, growth is carefully controlled by different plant growth regulators. Under these conditions, a bud continues to grow as part of a whole system. The tissue culturist excises the bud and alters the plant growth regulators. Growth then changes and multiple shoots will arise from a single bud. Each of these shoots has buds, which in turn can be cultured. The culturist can repeat this process over and over again to produce huge numbers of buds, each capable of producing a whole plant. Similar multiplication steps can be performed with rhizomes, nodes, tubers or other plant part. Micropropagation is the type of tissue culture used in the commercial horticulturist industry.

ORGANOGENESIS

Organogenesis refers to the development of plant organs such as leaves, roots or shoots from undifferentiated callus. A callus is a mass of "unorganized" plant cells. Callus can occur as a result of a plant wound. In plant tissue culture, it usually arises from pieces of tissue which the culturist places on a medium containing certain plant growth regulators. When the culturist moves the callus onto a new medium with different plant growth regulators, the cells in the callus are induced to organize and form new plant organs.

EMBRYOGENESIS

Plant embryos can arise from the manipulations of certain calluses. These embryos arise from undifferentiated cells in the callus. They reform into embryos. The term indirect embryogenesis is used when the embryos arise from a callus. If the embryos form from developed plant tissue, the term direct embryogenesis is used. Generally, embryogenesis is defined as a embryo arising from a single cell.

EMBRYO/OVULE CULTURE

In some plants, embryos or ovules have insufficient nutrient reserves to develop and germinate. Orchids are excellent examples. The seeds of orchids must be placed on a nutrient medium to allow full germination. This medium contains important salts, vitamins, sugar and plant growth regulators. The latter may improve germination or overcome any germination blockers.

PROTOPLAST CULTURE AND USES

When a plant cell is stripped of its wall, it is bound only by its plasmalemma. This cell is called a protoplast. Such cells may be fused with other cells to create entirely new species. This process is called protoplast fusion. The new fused protoplasts are cultured to reform new cell walls. The complete plant cells are then cultured back into new plants. Protoplasts can also be excellent choices for the injection of foreign DNA.

LABORATORY TECHNIQUES FOR PLANT TISSUE CULTURE

Proper tissue culture procedures require aseptic techniques. Those students who have taken microbiology are already familiar with these techniques. The culturist is the biggest factor in maintaining axenic cultures. Following good, sterile technique, the laminar flow hood represents a major breakthrough for tissue culture. These hoods filter the incoming air of all particles greater than 0.2 ìm in size. The hood directs a gentle (even unnoticeable) stream of sterile air out of the enclosure towards the culturist. This prevents air-borne particles from blowing into the work area. As the cultures are manipulated in the space of the hood, the airstream creates a sterile zone and the flow prevents airborne microbes from landing in the culture vessels or on the plants.

In practice, the culturist must perform all media and tissue preparation operations using sterile techniques so that foreign microbes are not introduced into the axenic cultures. With the high sucrose and nutrient content of the media, heterotrophic microbes would have a wonderful environment in which to grow, multiply and over-run the cultured plants. Good techniques call for:

1. Wearing clean clothing and keeping long hair away from the cultures.
2. Cleaning the laminar flow hoods prior to and after use.
3. Keeping fingers or other body parts downstream and away from the cultures. This requires that the vessels be manipulated carefully and in the sterile air flow.
4. Keeping the tools, i.e. scalpels, forceps, etc clean and sterile through frequent dips in 70% ethanol followed by heat treatments. These techniques will become clearer when your instructor demonstrates them to you. Cleanliness, good sterile technique, and attention to keeping your airway clear are of paramount importance for success in any plant tissue culture

Uses of plant tissue culture

Plant tissue culture now has direct commercial applications, as well as value in basic research into cell biology, genetics, and biochemistry. The techniques include culture of cells, anthers, ovules, and embryos on experimental to industrial scales, protoplast isolation and fusion, cell selection, and meristem and bud culture. Applications include:

1. micropropagation using meristem and shoot culture to produce large numbers of identical individuals

2. screening programs of cells, rather than plants for advantageous characters large scale growth of plant cells in liquid culture as a source of secondary products
3. crossing distantly related species by protoplast fusion and regeneration of the novel hybrid
4. production of diploid plants from haploid cultures to achieve homozygous lines more rapidly in breeding programs
5. As a tissue for transformation, followed by either short term testing of genetic constructs or regeneration of transgenic plants removal of viruses by propagation from meristematic tissues.

MEDIA

Since plant tissue culture involves the in vitro culture of plants or cells in small containers, careful attention must be paid to the plants' nutritional needs. In addition, small pieces of plants may not be able to manufacture all their own food through autotrophic means. The culturist usually needs to supply some other form of energy rich compound, such as sucrose. Also, the medium should contain the kinds of plant growth regulators which the culturist hopes will produce the desired forms of growth. Such growth might be rapid shoot multiplication, callus formation, organogenesis or embryogenesis.

Plant tissue culture media are made up of many components. These include inorganic salts, sucrose or some other carbon source, plant growth regulators and vitamins which the plants are often unable to make for themselves.

Another important feature of the different media is the pH. The pH affects both the growth of plants and the activity of the plant growth regulators. A pH of 5.7 is used for most plant tissue culture applications.

The culturist prepares media in several ways. The two most common approaches are to prepare stock solutions or buy the salts in premixed bottles.

The first step of the process is to measure out about 70-80% of the desired volume of distilled water needed. In other words, if the culturist needs 1 L of a medium, 700 to 800 mls of distilled water are poured into an appropriate flask. (A volume closer to 700 ml is used if liquid stock solutions provide the nutrients, while as much as 850 ml can be used if premixed salts provide the nutrients). The salts are dissolved in the water. The culturist then adds sucrose, usually around 1 to 3%. Both vitamins and plant growth regulators are added from stock solutions. After insuring that everything is dissolved, the culturist adjusts the pH (usually 5.7). The medium is then poured back into a graduated cylinder where it can be adjusted to the final volume.

The media may be supplied to the plants in a liquid form or it may be solidified with agar. Agar is a polysaccharide from marine red algae which produces a gel when dissolved in water and allowed to cool. In plant tissue culture, the agar content of a solid medium is around 0.8%. In order to dissolve the agar, the culturist heats the medium to boiling with the agar in it. After the medium becomes translucent, the agar is dissolved.

The medium is then dispensed into the containers. The individual containers with their lids are then autoclaved to sterilize the medium (121°C, 15 psi, 20 minutes). In some instances, plant growth regulators may not be autoclavable. In these cases, the plant growth regulators are filter sterilized (through a 0.2 ìm filter) and added to the medium after it is autoclaved. In such cases, the containers may be sterilized without the medium. When the fully complete medium is finished, it is then poured into the containers.

DEMONSTRATION OF TISSUE CULTURE FOR TEACHING

Introduction

The starting point for all tissue cultures is plant tissue, called an explant. It can be initiated from any part of a plant root, stem, petiole, leaf, or flower although the success of any one of these varies between species. It is essential that the surface of the explant is sterilized to remove all microbial contamination. Plant cell division is slow compared to the growth of bacteria and fungi, and even minor contaminants will easily overgrow the plant tissue culture. The explant is then incubated on a sterile nutrient medium to initiate the tissue culture. The composition of the growth medium

Table : Formulations of selected plant tissue culture media

Compound	Murashige Skoog	Woody Plant Medium	Anderson	White's	Gamborg B5	Heller
	concentration (mg/liter)					
NH_4NO_3	1650	400	400			
KNO_3	1900		480	80	2500	
$Ca\ (NO_3)_2\ 4H_20$		556		300		
$NaNO_3$						600
$(NH_4)_2SO_4$					134	
$MgSO_4\ 7H_2O$	370		370	720	250	
$NaSO_4$				200		
$CaCl_2\ 2H_2O$	440	96	440		150	75
KH_2PO_4	170	170	380			125
$NaH_2PO_4\ H_2$				16.5	150	
KCl				65		750
K_2SO_4		990				
$MgSO_4\ 7H_20$		370				250
$FeSO_4\ 7H_20$	27.8	27.8	55.7		27.5	
Na_2EDTA	37.3	37.3	74.5		37.3	
$FeCL_3\ 6H_20$						1
$Fe_2\ (SO_4)_3$				2.5		
$MnSO_4\ H_20$	16.8	22.3	16.9		10	0.01
$ZnSO_4\ 7H_20$	8.6	8.6	8.6	3	2.0	1
$H_3\ BO_3$	6.2	6.2	6.2	1.5	3.0	1
KI	0.83			0.75	0.75	0.01
$Na_2MoO_4\ 2H_20$	0.25	0.25	0.25		0.25	
$CuSO_4\ 4\ 5H_20$	0.025	0.25	0.025		0.025	0.03
$CoCl\ 6H_20$	0.025		0.025		0.025	
$NiCl\ 6H_20$						0.03
AlCL3						0.03
myo-inositol	100	100	100		100	
nicotinic acid	0.5	0.5		0.5	0.5	
pyridoxine HCl	0.5	0.5		0.1	1.0	
thiamine HC1	0.1	1.0		0.1	10.0	1.0
glycine	2.0	2.0		3.0		
Ca D-pantothenic acid				1.0		

is designed to sustain the plant cells, encourage cell division, and control development of either an undifferentiated cell mass, or particular plant organs.

The concentration of the growth regulators in the medium, namely auxin and cytokinin, seems to be the critical factor for determining whether a tissue culture is initiated, and how it subsequently develops. The explant should initially form a callus, from which it is possible to generate multiple embryos and then shoots, forming the basis for plant regeneration, and thus the technology of micropropagation. The first stage of tissue culture initiation is vital for information on what combination of media components will produce a friable, fast growing callus, or a green chlorophyllous callus, or embryo, root, or shoot formation.

There is at present no way to determine the exact growth medium, and growth protocol, to generate a particular type of callus. These characteristics have to be determined through a carefully designed and observed experiment for each new plant species, and frequently also for each new variety of the species that is taken into tissue culture. The basis of the experiment will be media and protocols that give the desired effect in other plant species, and experience.

The Demonstration

The strategy for designing a medium to initiate tissue culture, showing how growth regulators and other factors modulate development, can be demonstrated using the African violet, a popular house plant. Leaf sections are the source of explants. This demonstration is regularly carried out by a student class, and produces reliable results. Sterile supplies are provided from central facilities, and provision of sterile working areas (for example, in laminar flow hoods) is an advantage, although cultures can be initiated in an open laboratory with careful aseptic technique. The standard precautions used during any laboratory work involving chemicals or microbes should be adopted. If you are in any doubt about safety hazards associated with this demonstration, you should consult your local safety advisor.

Step 1. *Selection of the Leaves*

Leaves are cut from healthy plants, leaving a short length of petiole attached. They should be selected to each yield several explants of leaf squares, with approximately 1-cm sides. The youngest and oldest leaves should be avoided.

Wash the dust off the leaves in a beaker of distilled water, holding the leaf stalk with forceps.

Step 2. *Surface Sterilization and Preparation of the Explants*

This part of the procedure should be carried out in a sterile working area, or with meticulous aseptic technique.

The leaf, with the petiole still attached, should be immersed in 70% ethanol for 30 seconds, and then transferred to a sterile Petri dish. Sterile scissors and forceps are then used to cut squares from the leaf as explants, each with approximately 1 cm sides.

The explants are transferred into a 10% hypochlorite bleach solution for 5 minutes, gently agitating once or twice during this time. They are then washed free of bleach by immersing them in 4 successive beakers of sterile distilled water, leaving them for 2 to 3 minutes in each. Three explants are placed on each Petri dish of growth medium, with the upper epidermis pressed gently against the surface of the agar to make good contact.

The Petri dishes are sealed with plastic film to prevent moisture loss, and incubated at 25°C in 16 hrs is light/8 hrs dark.

Step 3. *Assessment of Tissue Culture Development*

The explants are incubated for 4 to 6 weeks, and inspected at weekly or biweekly intervals. The growth of obvious bacterial or fungal colonies indicates contamination and data from such cultures are obviously suspect. The development of dark brown tissue cultures can also be a consequence of contamination.

The media used in the demonstration are designed to show the effects of auxin, cytokinin, sucrose, and mineral salts on development. The media were based on the well known Murashige and Skoog inorganic media.

Typical Results

1. After about 8 weeks on each medium typical result appear. To summarize, multiple adventitious buds form on the control medium, leading to many small shoots on the upper surface where the leaf is not in contact with the medium.
2. Absence of sucrose inhibits this production. Shoot production is also limited on the low sucrose concentration, but comparable with the control at high sucrose.
3. At zero and low levels of cytokinin, callus forms where the leaf surface is in contact with the medium, while at high levels, shoot formation is stimulated.
4. At zero and low levels of auxin there is a stimulus to shoot formation, but at high concentrations, large numbers of roots are formed.
5. At low and zero levels of MS salts, there is no growth at all.

These very obvious variations demonstrate the importance of a carbon and inorganic salt source for plant growth, as well as the effect of the auxin: cytokinin ratio on the control of plant development.

PREPARATION OF PLANT TISSUE CULTURE MEDIA

Since the pioneer plant tissue culture studies, one of the main objectives was to design a proper medium that supports sufficient growth of the explants in a totally artificial environment. The first medium formulations used for plant culture work were based on experience of microorganisms. These media contained several distinct classes of compounds:

1. Carbon source
2. Organic supplements
3. Inorganic salts
4. Trace elements.

As works on plant cell cultures progressed, attention was paid in order to specifically define media for an optimal growth of the variable plant cells, tissue, and organs. Today, the most commonly used culture media are based on following components:

1. Macroelements, N, P, K, Ca, Mg, and S.
2. Microelements, Fe, Cu, Mn, Co, Mo, B, I, Zn, Cl and sometimes Al, Ni, and Si.
3. Carbon source, Sucrose, glucose, fructose, or sorbitol.
4. Organic compounds (Vitamins), thiamine, niacin, pyridoxine, biotin, folic acid, ascorbic acid, tocopherol.
5. Myoinositol or casein hydrolysate.
6. Complex organics Coconut milk and water, yeast extract, fruit juices, or pulps.
7. Plant growth regulators Auxins, cytokinins, Gibberellins, abscisic acid.
8. Gelling agents Agar, agarose, gellan gum.
9. Other components MES, activated charcoal, antibiotics, fungicides.

The number of plant species that have been cultured and the different media used are extensive. Once the plant species to be cultured is decided upon, a suitable medium formulation should be selected. One of the most successful media, is the MS medium, was devised by Murashige and Skoog (1962), based on the constitution of tobacco plants. The choice or formulation of the media basically depends on the requirements of the cultivated explant and species.

Media Formulation

As mentioned above, MS medium was developed for culture of tobacco explants and the formulation was based on the analysis of the mineral compounds present in the tobacco tissue itself. It is noted for its high salt levels, particularly K and N salts. LM medium (Linsmaier and Skoog, 1965) is a revision of the MS medium, in which the organic constituents have been modified and only thiamine and myo-inositol retained. White's medium (White, 1963) was devised for the culture of tomato roots and has lower concentrations of salts than MS medium. Gamborg's B5 medium (Gamborg et al., 1968) was devised for soybean callus culture and contains a much greater proportion of nitrate compared to the ammonium. SH medium (Schenk and Hildebrandt, 1972) was developed for callus culture of both monocotyledons and dicotyledons. Nitsch's medium (Nitsch and Nitsch, 1969) was developed for another culture. It contains lower salt concentrations than that of MS, but not as low as that of White's medium. KM medium (Kao and Michayluck, 1975) was designed to grow cells and protoplasts of Vicia at a very low cell density in liquid media. The medium of Chu (N6) is defined to improve the formation, growth, and differentiation of pollen in rice. The concentration of ammonium proved to be crucial for the development of callus. Recently other media have been developed for specific explants and species.

PROTOPLAST ISOLATION, CULTURE, AND FUSION

Background information

Protoplasts are cells that have had their cell wall removed, usually by digestion with enzymes. Cellulase enzymes digest the cellulose in plant cell walls while pectinase enzymes break down the pectin holding cells together. Once the cell wall has been removed, the resulting protoplast is spherical in shape.

Digestion is usually carried out after incubation in an osmoticum (a solution of higher concentration than the cell contents that causes the cells to plasmolyze). This makes the cell walls easier to digest. Debris is filtered and/or centrifuged out of the suspension and the protoplasts are then centrifuged to form a pellet. On resuspension, the protoplasts can be cultured on media that induce cell division and differentiation. A large number of plants can be regenerated from a single experiment a gram of potato leaf tissue can produce more than a million protoplasts, for example.

Protoplasts can be isolated from a range of plant tissues: leaves, stems, roots, flowers, anthers, and even pollen. The isolation and culture media used vary with the species and with the tissue from which the protoplasts were isolated.

Protoplasts are used in a number of ways for research and plant improvement. They can be treated in a variety of ways (electroporation, incubation with bacteria, heat shock, high pH treatment) to induce them to take up DNA. The protoplasts can then be cultured and plants regenerated. In this way, genetically engineered plants can be produced more easily than is possible using intact cells/plants.

Plants from distantly related or unrelated species are unable to reproduce sexually, as their genomes/modes of reproduction, etc. are incompatible. Protoplasts from unrelated species can be fused to produce plants combining desirable characteristics such as disease resistance, good flavour and cold tolerance. Fusion is carried out by application of an electric current or by treatment with chemicals such as polyethylene glycol (PEG). Fusion products can be selected for on media containing antibiotics or herbicides. These can then be induced to form shoots and roots and hybrid plants can be tested for desirable characteristics.

Materials

1. Lettuce (round, green lettuce, not iceberg)
2. Scalpel or sharp knife
3. Forceps
4. Tile
5. Glass Petri dish
6. 10 cm^3 syringe
7. 1 cm^3 syringe 13% sorbitol solution
8. Viscozyme enzyme
9. Cellulase enzyme
10. Dropper
11. Small filter funnel
12. 60 mm gauze square (approximately 12 cm × 12 cm)
13. Tape
14. Centrifuge tube
15. Slide and cover slip

A centrifuge, high-power microscope, and incubator set at approximately 35°C must be available in the lab.

Preparation of Materials

It is best to use limp lettuce that has been left in an incubator at about 35°C for an hour. This causes the cells to plasmolyze slightly.

Do not dilute the enzymes. The enzyme solution should be stored in a fridge until use.

To prepare a 13% sorbitol solution, add 18.5 cm^3 stock to 100 cm^3 distilled water.

Waterproof sticky tape can be used to secure the gauze mesh in a small filter funnel. This can be washed and reused.

For Protoplast Fusion

A microcentrifuge can be used to spin the protoplasts. A spin of 3 minutes at the lowest voltage is sufficient. Care should be taken when resuspending the Pellet, as the protoplasts are very fragile.

Making the Fusion Mix

1. PEG (12 ml of 50% solution)
2. HEPES buffer 0.9 g
3. 10 ml water
4. pH 8.0 Materials
5. Lettuce leaf (2 squares, each approximately 3 cm × 3 cm)
6. Scalpel or sharp knife
7. Forceps
8. Tile
9. Glass Petri dish
10. 10 cm^3 syringe
11. 1 cm^3 syringe
12. 13% sorbitol solution
13. 1.2 cm^3 viscozyme enzyme
14. 0.6 cm^3 cellulase enzyme

15. Dropper
16. Small-filter funnel
17. 60 mm gauze
18. Tape
19. Centrifuge tube
20. Slide and cover slip
21. Centrifuge
22. Incubator
23. Microscope

Procedure

1. Place 12.5 cm3 of 13% sorbitol solution into the Petri dish.
2. Place the squares of lettuce leaf bottom side in the Petri dish.
3. Cut up the lettuce leaf into pieces approximately 5 mm × 5 mm.
4. Place the lid on the dish and incubate in the oven at 35°C for 5 minutes.
5. Add the viscozyme and cellulase enzymes to the dish. Swirl gently.
6. Incubate for a further 20 minutes, swirl gently at regular intervals.
7. Place the 60-mm mesh into the funnel and secure with tape.
8. Hold the funnel over the lid of the Petri dish.
9. Pour the digested lettuce into the filter funnel. Try to minimize the distance the protoplasts have to fall.
10. Place the funnel over a centrifuge tube and gently pour the contents of the Petri dish lid through the funnel again.
11. Balance your tube with another and centrifuge for 5 to 10 minutes at 2000 rpm.
12. The protoplasts will have formed a pellet at the bottom of the tube. Use the dropper to remove the liquid supernatant from the tube, without disturbing the pellet. Remove all but 0.5 cm^3 of the liquid and then resuspend the pellet by gently tapping the tube.
13. Place a drop of the suspension on a slide, cover with a cover slip, and examine under high power.

14. If your suspension is too concentrated to see the protoplasts, clearly you can add several drops of the sucrose solution to the suspension in the centrifuge tube.
15. You should be able to try to fuse your protoplasts together if you have time.

Protoplast Fusion (Somatic Hybridization)

1. Glass slide
2. Cover slip
3. Glass dropper
4. Fusion mix
5. Protoplast suspension
6. Microscope

Instructions

1. Place a drop of the protoplast suspension on the glass slide.
2. Observe under the medium power of the microscope.
3. Add 1 drop of the fusion mix.
4. Observe over a period of ~3 minutes.

AGROBACTERIUM CULTURE AND AGROBACTERIUM MEDIATED TRANSFORMATION

Materials

1. Infiltration media selection plates
2. 5X MS salts 1X MS salts
3. 1X B5 vitamins .3% sucrose
4. 5% sucrose 8 g TC agar
5. 044 mM benzylamino purine
6. 03% Silwet L-77

Autoclave and add appropriate antibiotics (for pCGN add Kan 50 mg/ml, timentin 200 mg/ml).

Procedure

Plant Growth

1. Plant seeds of appropriate genotype on top of cheese cloth-covered soil, making sure the soil fill the pot. Use a rubber band to secure the cheesecloth.
2. After plants have bolted, clip off the primary bolt to encourage the growth of secondary bolts. Perform infiltration 4-8 days after clipping. Vacuum infiltration 1. Grow a large liquid culture of Agrobacterium (such as ASE strain) carrying the appropriate construct. Start a 25 ml overnight (LB+ antibiotics) 2 to 3 days ahead of infiltration. One day prior to infiltration, use this culture to inoculate a 400-mL culture.
3. After 24 hours of growth, cells are usually at a density of at least 1 OD. Harvest cells by centrifugation and resuspend to an OD of .8 in infiltration media.
4. Use a crystallization dish for the actual infiltration but the pots do not fit exactly, so any ideas/improvements in this area are encouraged. Invert the pot and stick it into the infiltration solution. Put into the vacuum oven at RT. Infiltrate for 10 to 15 min at 10 to 15 in 3 Hg.
5. Release vacuum and remove pot, lay it on its side in the tray, and cover it with Saran wrap. Remove the cover the next day and stand upright.
6. The Agro infiltration solution can be reused for multiple pots.

Selection of Transformants

7. Pour selection plates.
8. Sterilize seeds in 50% water/50% bleach/drop of Tween. Rinse 3–4 times with sterile water.
9. Plate seeds by resuspending in room temperature .1% agarose and Pipetting onto selection plates. Dry plates in laminar flow hood until set. Use 1 ml agarose for every 500 to 1000 seeds. Plate 2000 to 4000 seeds per 150 × 15 mm plate. Put a small number of control seeds from a known transformed plant on one of the plates for comparison.

10. Vernalize plants for 2 nights in cold room. Move plates to growth chamber.
11. After about 7 days, transformants should be clearly visible as dark green plants with roots that extend over and into the selection media.
12. Transfer seedlings to soil.

ISOLATION OF CHLOROPLASTS FROM SPINACH LEAVES

Procedure

1. Prepare an ice bath and precool all glassware to be used (including a mortar and pestle).
2. Select several fresh spinach leaves. Remove the large veins by tearing them loose from the leaves. Weigh out 4 g of deveined leaf tissue.
3. Chop the tissue as fine as possible with a knife and chopping board.
4. Add the tissue to an ice cold mortar containing 15 ml of grinding solution and grind the tissue to a paste.
5. Filter the ground up tissue solution through double layered cheesecloth into a beaker, and squeeze the tissue pulp to recover all of the liquid suspension.
6. Transfer the green suspension to a cold 50-mL centrifuge tube and centrifuge for 1 minute at 200 xg at 4°C to pellet the unbroken cells and cell fragments.
7. Decant the supernatant into a clean centrifuge tube and centrifuge again for 7 minutes at 1000 xg to pellet the chloroplasts. Decant and discard the supernatant.
8. Resuspend the chloroplasts in 5 ml of cold suspension solution (or 0.35 M NaCl). Use a cold glass stirring rod to gently disrupt the packed pellet.
9. Enclose the tube in aluminum foil and place at 0°C to 4°C.
10. Using a hemacytometer, determine the number of chloroplasts per ml of suspension media.

Solutions

1. Spinach leaves (fresh) suspension solution
2. 0.33 M Sorbitol
3. pH 7.6, adjust with NaOH
4. 1 mM MgCl2
5. 50 mM HEPES
6. 2 mM EDTA

Grinding Solution

1. 0.33 M Sorbitol
2. pH 6.5, adjust with HCl
3. 4 mM MgCl2
4. 2 mM ascorbic acid (Vitamin C)
5. 10 mM sodium pyrophosphate
6. 0.35 M NaCl

Bioreagents and Chemicals

1. Sodium hydroxide
2. Sodium pyrophosphate
3. Sodium chloride
4. Spinach leaves (fresh)
5. Magnesium chloride
6. Ascorbic acid
7. Sorbitol
8. HEPES
9. Cheesecloth
10. Hydrochloric acid
11. EDTA

Protocol

The isolation procedure used in this protocol leaves the chloroplast outer membrane intact. If you wish to study the enzymes for photophosphorylation, wash the chloroplasts and rupture the outer membranes. To rupture the outer membranes, resuspend the chloroplasts in diluted suspension solution (1:25). Immediately centrifuge the chloroplast suspension for 5 minutes at 8000 xg to collect the chloroplasts. Remove the diluted suspension media and resuspend the chloroplasts in isotonic media (0.35 M NaCl or undiluted suspension buffer).

PROTOCOLS FOR PLANT TISSUE CULTURE

Tissue culture involves several stages of plant growth, callus formation, shooting, roots, and flowering. It starts with sterilizing a piece of plant tissue.

Sterilizing Plant Materials

Inside the glove box, sterilize all plant material in diluted domestic bleach solution containing 1 drop of detergent as a surfactant. Put plant pieces in a jar containing the bleach for 10–20 minutes. Agitate frequently. Transfer the tissue in the bleach solution to the sterile area. Discard the chlorine solution. Rinse plant pieces twice with sterile water.

(A) Callus Formation Protocol

1. Prepare a batch of basic media without added coconut milk or malt extract.
2. Follow the sterilization steps for the medium, instruments, and chamber.
3. Organize the cultures, sterile media, sterilized tools, and sterile paper towels, and sterile water at one end of the chamber. At the other end of the chamber, organize a small paper bag for trash, a jar of bleach solution, and a spot for freshly inoculated cultures. The idea is for materials to always move in the same direction. This will help keep you from getting confused.
4. Sterilize your plant tissue, following this protocol.
5. Take a sterilized piece of tissue from the jar with a pair of forceps (do not touch the plant material with your hands). Working on a dampened sterile paper towel, use the scalpel or a razor blade to cut the tissue sample into 2 to 3 cm long pieces.

6. Put 1 piece of the tissue into each container (it is important to have only 1 shoot per container at this stage so that if the shoot is contaminated it cannot spread to the others). Shut the lids of the containers.
7. Store jars at room temperature away from direct sunlight. Callus will start to become visible after about 7 days post-initiation.
8. Friable type II callus should be separated from more organized callus and/ or watery unorganized callus 2 to 3 weeks after initiation.
9. Friable type II callus should be visually selected at each subsequent transfer to maintain an optimal phenotype.
10. Callus should be transferred to fresh medium at 2 to 4week intervals, depending on growth rate. Switching to different types of media also helps maintain vigor.
11. Callus can be maintained at room temperature in the dark.

(B) Shoot Multiplication Protocol

1. Prepare a batch of media containing 5% to 10% coconut milk.
2. Follow the sterilization steps for the medium, instruments, and chamber. Organize the cultures, sterile media, sterilized tools, sterile paper towels, and sterile water at one end of the chamber. At the other end of the chamber organize a small paper bag for trash, a jar of bleach solution, and a spot for freshly inoculated cultures. The idea is for materials to always move in the same direction. This will help keep you from getting confused.
3. With a pair of forceps, remove a culture from its container. Moisten the sterile paper with some sterile water. It is important to do all the manipulation on a damp paper towel, as these plants are very soft and can desiccate readily.
4. Cut stems off the culture and transfer to new jars. At this stage, up to 5 stems may be put inside each jar.
5. Store cultures as explained in the previous stage.

(C) Root Formation Protocol

Once you have established enough shoots, let them grow to at least 2 cm before beginning the rooting process.

1. Prepare a batch of media containing malt extract.
2. Follow the sterilization steps for the medium, instruments, and chamber.
3. Organize the cultures, sterile media, sterilized tools, sterile paper towels, and sterile water at one end of the chamber. At the other end of the chamber, organize a small paper bag for trash, a jar of bleach solution, and a spot for freshly inoculated cultures. The idea is for materials to always move in the same direction. This will help keep you from getting confused.
4. With a pair of forceps, remove a stem from its container. Wash all of the media off the culture and transfer to malt media. (Malt contains auxins, which promote root formation.)
5. Up to 5 shoots may be put in each culture vessel. Store containers in their usual place as before. Roots should form within 2 to 4 weeks.

(D) Acclimatization Protocol

1. Fill plastic "vegetable" bags with a potting mix that contains no fertilizer.
2. Autoclave for 15 minutes in the pressure cooker.
3. When cool, inoculate with mycorrhizae.
4. Remove the rooted plants from agar medium using a pair of forceps.
5. Wash off the agar thoroughly from the roots using lukewarm water.
6. Poke a hole in the middle of the potting mix, using a sterile instrument, gently insert the roots in that hole.
7. Dampen the potting mix with basic nutrient mix.
8. Spray the foliage with a hand spray containing sterile water.
9. Keep these bags inside larger plastic containers with a glass cover, out of direct sunlight. Gradually remove the glass cover, but watch for signs of desiccation and if needed, use the hand spray to spray water on the foliage. Gradually increase the light intensity for the plants also.
10. When the roots are well established and the plants are acclimatized (this should take about 4 to 6 weeks), they can be given fertilizer and be treated like any other plant.

(E) Flowering Protocol

1. Prepare a batch of jars with peroxidated media.
2. Follow the sterilization steps for the instruments and chamber.
3. Organize the cultures, sterile media, sterilized tools, sterile paper towels, and sterile water at one end of the chamber. At the other end of the chamber, organize a small paper bag for trash, a jar of bleach solution, and a spot for freshly inoculated cultures. The idea is for materials to always move in the same direction. This will help keep you from getting confused. With a pair of forceps, remove a plantlet from its container. Wash all of the media off the plantlet and transfer it to peroxidated media.
4. Watch the cultures carefully for growth of flower buds.

(F) Anther/Ovule Culture Protocol (For Producing Haploid Lines)

1. Collect immature flower buds from plants grown either in the field, the greenhouse, or in vitro.
2. Preculture the flower buds in dry test tubes in the refrigerator for 2 days.
3. Dip the flower buds in 70% ethanol for 30 seconds, and then surface sterilize with 0.5% to 1.0% sodium hypochlorite for 20 min in your glove box.
4. Rinse the flower buds with sterile distilled water 3 times, and aseptically excise the anthers and/or ovules using a scalpel and a needle.
5. Culture the anthers/ovules on semisolid nutrient medium.
6. Incubate the cultures under complete darkness at room temperature.
7. Once calli or embryos are initiated, transfer the cultures to daylight.
8. Culture the developing embryoids on standard propagation medium.

(G) Plant Protoplast Preparation Protocol

1. Perform all of the following in the glove box.
2. **Buffer Solution :** Dissolve 56.94 g of mannitol in 200 ml of distilled water. Add distilled water to bring the final volume to 500 ml.

3. **Digestion Solution :** Measure 10 ml of buffer solution into a 15 ml test tube or small beaker. Measure 0.1 g pectinase and 0.2 g cellulace onto weighing paper. Drop the premeasured, powdered enzymes into this solution. Swirl the beaker or cap the test tube and shake it back and forth until the enzymes are completely dissolved.
4. Carefully pour all 10 ml of enzyme solution into the bottom of a sterile jar.
5. Use forceps to float each tissue sample on the surface of the enzyme solution.
6. Seal the jar with Saran Wrap or tape and leave it at room temperature (approximately 25°C) overnight. If proper equipment is available, gentle agitation of the dishes will be helpful.
7. The next day, gently swirl and shake the solution in the Petri dish. (If no protoplasts are observed with the microscope, let the solution stand for another 15 to 30 minutes, then look again for protoplasts.)
8. At the end of the digestion period, gently shake the Petri dish to release the protoplasts.
9. Filter the enzyme-protoplast suspension through successively smaller filters, starting with the 100-mm sieve. One sieve will trap mostly protoplasts.

(H) Chemical Protoplast Fusion Protocol

1. Mix 2 droplets of protoplasts from 2 genetically different strains, along with a droplet of PEG (polyethylene glycol) in a test tube.
2. Flood the protoplasts with baking soda dissolved in basic growing mix.
3. After 20 minutes, wash with basic growing mix and centrifuge in the salad spinner centrifuge.
4. Transfer the fused protoplasts onto callus culture medium.

(I) Electrical Protoplast Fusion Protocol

1. Place 2 droplets of protoplasts from 2 genetically different strains, between 2 electrodes.
2. Apply 10 volts per millimeter of electrode separation for 1 minute.

3. Apply a 10 to 20 msec pulse at 100 volts per millimeter. This will require building an electrical pulse apparatus. A 555 timer discharging through an automotive coil might do the trick.
4. Wash with basic growing mix and centrifuge in the salad spinner centrifuge.
5. Transfer the fused protoplasts onto callus culture medium.

(J) Polyploidization Protocol

1. Prepare a batch of media containing 0.001% to 0.01% trifluralin (Preen or Treflan brand name).
2. Follow the sterilization steps for the medium, instruments, and chamber.
3. Organize the cultures, sterile media, sterilized tools, sterile paper towels, sterile water, and bleach solution in the glove box.
4. Place a clean culture or fused protoplast on trifluralin media.
5. Store containers in a cool, dark location. Within 2 to 4 weeks, polyploidy shoots should begin developing.

APPENDIX—A

UNITS AND MEASURES

A. Size

Cell biology deals with things that are relatively small. The units of measurement typically used are the micron at the light microscope level, and the nanometer at the electron microscope level. For molecular measurements, the norm is the Angstrom (Å). These units are defined within the following table:

Measure	Symbol	Relative Length	Exponential Notation
Meter	M	1	100
Decimeter	dm	1	10–1
Centimeter	cm	.01	10–2
Millimeter	mm	.001	10–3
Micrometer or micron	mm	.000001	10–6
Nanometer	nm	.000000001	10–9
Angstrom	Å	.0000000001	10–10

From this table it is apparent that:

10 Å = 1 nm

1000 nm = 1 mm

10 mm = 1 cm

Not apparent is that:

1 inch = 2.54 cm = 25.4 mm = 25400 μm = 25400000 nm

1 inch = 2.54 cm = 2.54×10^1 mm = 2.54×10^4 mm = 2.54×10^7 nm

1 mm = 0.04 inches.

B. Basic Units Used in Biochemistry

1. Mole Units

1 mole

1 millimole (mmole) = 10^{-3} moles

1 micromole (mmole) = 10^{-6} moles

1 nanomole (nmole) = 10^{-9} moles

1 picomole (pmole) = 10^{-12} moles

2. Gram Units

1 gram (g)

1 milligram (mg) = 10^{-3} g

1 microgram (μg) = 10^{-6} g

Concentration = weight/volume

1 M (molar) = 1 mole/liter or 1 mmole/ml

1 mM = 1 mmole/liter, etc.

3. Volume : Volumes are measured relative to a liter, with the most commonly used measurements, the milliliter, and the microliter. The following table provides the relative volumes:

Measure	Symbol	Relative Volume	Exponential Notation
Liter	L	1	10^7
Deciliter	dL	.1	10^{-1}
Milliliter	Ml	.001	10^{-3}
Microliter	μl	.000001	10^{-6}

There are 1000 μl in 1 ml.

1 ml = 1 cm^3

1 gallon = 3.8 liters

1 quart = 0.95 liters

1 liquid ounce = 29.6 ml

4. **Weight :** The most common measurements of weight are the gram, milligram, and microgram.

Measure	Symbol	Relative Weight	Exponential Notation
Kilogram	Kg	1000	10^3
Gram	g	1	10^0
Milligram	mg	.0001	10^{-3}
Microgram	μg	.0000001	10^{-6}

5. **Concentration :** Most concentrations used throughout cell biology arc those of a solute dissolved or suspended within a solvent, and in most cases the solvent is water. There are 2 general methods of identifying the concentration of a solution; as molarity or as a percent. Molarity is based on the number of moles of solute in the solvent, while percent is based on the number of parts, either grams (for a solid solute) or milliliters (for a liquid solute).

   Molarity equals the number of moles of solute in 1 liter of solution. A mole is equal to the gram molecular weight (or formula weight) of the solute. Sodium chloride (NaCl), for example, has a formula weight of 58.43 (22.98 for Na and 35.43 for Cl). Thus, if 58.43 grams of NaCl are dissolved in 1 liter of water, the result would equal a 1molar solution of NaCl. This is designated as 1 M NaCl, or as simply M NaCl.

   We often deal with solutions of less volume than 1 liter, and the following should be noted :

   1 M NaCl = 58.43 grams/liter = 58.43 mg/ml.

   A 0.002 M NaCl solution contains 0.002 moles of NaCl or 0.1168 grams (0.002 × 58.43) in 1 liter of solvent. Note that molar is abbreviated as M, but that there is no abbreviation for moles. The 0.002 M solution contains 0.002 moles (or 2 millimoles) or solute in 1 liter. A 0.002 M solution would contain moles of solute in a half-liter.

   The number of moles = Volume (in liters) × Molar concentration

   The number of millimoles = Volume (in ml0) × Molar concentration

Note that chemical equations are always balanced via moles. Moreover, note that for dilutions of known concentrations, one can use the simple formula:

$$\text{Molarity} \times \text{Volume} = \text{Molarity} \times \text{Volume}$$

If you have a 0.002 M solution of NaCl and you wish to obtain 100 ml of a 0.001 M solution,

$$0.002 \text{ M} \times \text{Needed Volume} = 0.001 \text{ M} \times 100 \text{ ml}$$

$$\text{Needed Volume} = 0.001 \text{ M} \times .1 \text{ liters}/0.002 \text{ M} = .050 \text{ liters} = 50 \text{ ml.}$$

Measure 50 ml of the 0.002 M solution, and dilute it to 100 ml with the solvent (usually water, or an appropriate buffer).

Molarity is appropriate for use when chemical equations are to be balanced. When we deal with physical properties of solutions, molarity is not as valuable as a similar measurement of concentration, molality. For colligative properties of solutions (freezing point depression, boiling point elevation, osmotic pressure, density, viscosity), there is a better correlation between the property and molality.

Molality (designated with a lower case m) is equal to the number of moles of solute in 1000 gm of solvent. At first, this may not appear any different from molarity, since an ml of water equals 1.0 gm. Indeed, for dilute solutions in water, there is little or no practical difference between a molar solution and a molal solution. In concentrated solutions, with temperature fluctuations and changes in solvent, there is appreciable difference.

For example, a 2 m (2 molal) solution of sucrose contains 684.4 gm of sucrose (twice the molecular weight or 2 moles of sucrose) dissolved in 1000 gm (approximately 1 liter) of water. The weight of this solution is 684.4 gm +1000 gm or 1684.4 gm.

This solution (2 m sucrose) has a density of 1.18 gm/ml or 1180 gm/liter. Since there are 1684.4 gm, division by the density (1180 gm/liter) would indicate that there are 1.43 liters of solution. That is, 684.4 gm of sucrose dissolved in 1000 gm of water would yield 1.43 liters of solution. This solution would contain 2 moles of sucrose, however, and would have a molarity = 2 moles/1.43 liters or 1.40 M. So, a 2 m sucrose solution equals a 1.4 M sucrose solution.

C. Percent (%) Solutions

In the example above of 2 m sucrose, there were 684.4 gm of sucrose in the final solution, which weighed 1684.4 gm (684.4 gm sucrose + 1000 gm water). The percent of sucrose on the basis of weight is therefore 684.4/1684.4 × 100, or 40.6%.

There are 3 means of expressing concentration in the form of a percent figure :

1. Percent by weight (w/w); gm solute/100 gm solvent
2. Percent weight by volume (w/v); gm solute/100 ml solvent
3. Percent by volume (v/v); ml solute/100 ml solution.

For dilute solutions, these differences are not significant, but at higher concentrations, they are. Chemists (when they use percent designations) usually use w/w. Biochemists and physiologists more often use w/v. Both use v/v if the solute is a liquid. It is important to distinguish between these alternatives.

Using ethanol as an example, consider a 20% solution of ethanol in water, mixed according to the 3 designations of w/w, w/v, and v/v.

1. w/w would contain 20 g of absolute ethanol mixed with 80 gm of water to yield a 20% (w/w) solution.
2. w/v would contain 20 g of absolute ethanol mixed with water to form a final volume of 100 ml.
3. v/v would contain 20 ml of absolute ethanol diluted to 100 ml with water.

The 3 solutions are not the same. First, the density of alcohol is not equal to that of water, and thus, conversion of g to ml is not equivalent. A 20% (w/w) solution of ethanol, for example, has a density of 0.97 g/ml and 20 gm of ethanol plus 80 gm of water would have a volume of 103 ml. The % (w/v) for this solution would be 20 gm ethanol/103 ml, or 19.4% (w/v). Similarly, absolute ethanol has a density of 0.79 gm/ml and thus 20 ml of ethanol would weigh 15.8 gm. A 20% (v/v) solution would contain 15.8 gm of ethanol in 100 ml and be a 15.8% (w/v) solution.

So, for ethanol:

$$20\%\ (w/w) = 19.4\ \%\ (w/v)$$
$$20\%\ (w/v) = 20.0\ \%\ (w/v)$$
$$20\%\ (v/v) = 15.8\ \%\ (w/v)$$

In cell biology, the most common use of percent solution is as (w/v). In practice, these are simple solutions to mix. For a 20% (w/v) sucrose solution, for example, simply weigh 20 gm of sucrose and dissolve to 100 ml with water.

Unless specifically stated otherwise, solutions lacking the appropriate designation should be assumed to be (w/v).

D. Dilutions

Examples

A. 1 ml of bacteria is mixed with 1 ml of sterile saline. The total ml in the tube would be 2 ml, of which 1 ml is bacteria. This is a 1:2 dilution (also written as ½, meaning ½ as many bacteria per ml as the original ml).

B. 1 ml of bacteria is mixed with 3 ml of sterile saline. The total ml in the tube would be 4 ml, of which 1 ml is bacteria. This is then a 1:4 dilution (also written ¼, meaning ¼ as many bacteria per ml as the original ml).

C. 1 ml of bacteria is mixed with 9 ml of sterile saline. The total ml in the tube would be 10 ml, of which 1 ml is bacteria. This is then a 1:10 dilution (also written 1/10 or 10^{-1}, meaning 1/10 or 10^{-1} as many bacteria per ml as the original ml).

D. For dilutions greater than 1:10, usually serial dilutions (dilutions of dilutions) are made. The following represents a serial 10 fold dilution (a series of 1:10 dilutions) :

   1. The dilution in tube 1 would be 1/10 or 10^{-1}.
   2. The dilution in tube 2 would be 1/100 or 10^{-2} (1/10 of 1/10).
   3. The dilution in tube 3 would be 1/1000 or 10^{-3} (1/10 of 1/100).
   4. The dilution in tube 4 would be 1/10000 or 10^{-4} (1/10 of 1/1000).

The dilution factor is the inverse of the dilution. (Inverse means you flip the 2 numbers of the fraction; with scientific notation you use the positive exponent).

1. For a dilution of 1/2, the dilution factor would be 2/1 or 2.
2. For a dilution of 1/4, the dilution factor would be 4/1 or 4.
3. For a dilution of 1/10 or 10^{-1}, the dilution factor would be 10/1 or 10 or 101.
4. For a dilution of 1/1000000 or 10^{-6}, the dilution factor would be 1000000/1 or 1000000 or 10^{6}.

In other words, the dilution factor tells you what whole number you have to multiply the dilution by to get back to the original 1 ml.

APPENDIX–B

CHEMICAL PREPARATIONS CHEMICALS OR EAGENTS REQUIRED FOR MOLECULAR BIOLOGY EXPERIMENTS

The following is a list of the solutions and chemicals required throughout the laboratory manual. It is organized alphabetically, and individual exercises list the materials needed for that exercise. For many solutions, directions are given for a molar solution and the user is left to dilute to the appropriate concentration for their needs. There are many vendors of the chemicals listed, and many of the solutions can also be purchased premixed.

Acetic Acid (MW 60.05)

Glacial acetic acid is 99.6% (w/v) acetic acid, and is 17.4 M.

- **a** **1 M :** Add 57.5 ml of glacial acetic acid to 800 ml of water and then make up to 1 liter with water.
- **b** **0.05 N :** Add 2.87 ml of glacial acetic acid to 800 ml of water and then make up to 1 liter with water.
- **c** **0.9 M :** Add 54 ml of glacial acetic acid to 800 ml of water and then make up to 1 liter with water.
- **d** **7% (w/v) :** Add 70 ml of glacial acetic acid to 800 ml of water and then make up to 1 liter with water.
- **e** **45% (w/v) :** Add 450 ml of glacial acetic acid to 500 ml of water and then make up to 1 liter with water.

Acetic Acid/Butanol/Water (15:60:25)

Combine 150 ml of glacial acetic acid, 600 ml of butanol, and 250 ml of water.

Acid Alcohol

Add 1.0 ml of concentrated HCl to 100 ml of 70% (v/v) ethyl alcohol.

Acrylamide Solutions

Acrylamide solutions for PAGE are given as total concentration of acrylamide (acrylamide + bis-acrylamide) and the amount of cross linker (bis-acrylamide). This is listed as the T:C ratio.

For example, a 10% gel (10%T:5%C) would contain a total of 10 grams of acrylamide per 100 ml, and would be composed of 5 grams of acrylamide and 5 grams of bis-acrylamide. Usually, a stock solution of 30% acrylamide is produced containing 0.8% bis-acrylamide. Many investigators use 30 grams of acrylamide plus 0.8 grams of bis-acrylamide per 100 ml of water, but 29.2 grams of acrylamide plus 0.8 grams of bis would be technically correct. In practice, it makes little difference since the gels are diluted to 10% or less. The 30% stock solution is filtered through a 0.45-m filter and stored at 4°C in the dark. For use, the stock solution is diluted with an appropriate buffer (usually a 2X Tris-HCl). The stock solution is stable for about 1 month. Discard after this period.

Acrylamides in their monomeric form are neurotoxic. Polymerize all acrylamide solutions prior to disposal.

SDS, β-mercaptoethanol and a tracker dye (bromophenol blue) are added at various points.

Alkaline Distilled Water

Add 1 pellet of NaOH to 1 liter of distilled water.

Alkaline Solution for G-banding

Dissolve 2.8 grams of NaOH and 6.2 grams of NaCl to a final volume of 1 liter with water.

Ammonium Acetate (MW 77.08)

0.1 M : Add 7.708 grams of Ammonium acetate to a final volume of 1 liter of water.

Ammonium Persulfate (MW 228.2)

10% (w/v) : Dissolve 1.0 grams of ammonium persulfate to a final volume of 10 ml with water. Mix fresh, prior to use as a catalyst for PAGE. Normally, about 50 ml of ammonium persulfate is added to each 15 of gel solution for polymerization. Dissolve 0.13 grams of Alcian Blue 8GX (Sigma- A-2899) in 100 ml of water.

Ammonium Sulfate (MW 132.14)

2% (w/v) : Add 2 grams of ammonium sulfate to a final volume of 100 ml water. M: Dissolve 542 grams of ammonium sulfate to a final volume of 1 liter.

Ascorbic Acid (MW 176.12)

2 mM : Dissolve 35.2 mg of ascorbic acid to a final volume of 100 ml with water.

ATP (Adenosine Triphosphate, MW 507.21)

5 mM : Dissolve 254 mg of ATP to a final volume of 100 ml with water or buffer. Dissolve 35.2 mg of ascorbic acid to a final volume of 100 ml with water.

Bis-acrylamide (N,N'-Methylene-bis-acrylamide)

Cross-linker for acrylamide gels.

Bovine Serum Albumin (BSA)

There are many grades of BSA available, and care should be taken when using this protein. For routine protein concentration standards, a 96% to

99% pure fraction (Sigma A 2153) may be used. For tissue culture, RIA, or molecular weight standardization, BSA should be obtained which is extracted and purified specifically for that purpose.

1% (w/v) : Dissolve 0.5 grams of BSA to a final volume of 50 ml in water or buffer.

Bradford Protein Assay

This procedure uses an absorbance shift in an acidic Coomassie Blue solution. It is commercially available from Pierce Chemical Company, Rockford, Illinois as Protein in Assay Reagent. It contains methanol and solubilizing agents and is very reliable.

If you wish to make your own, dissolve 100 mg of Coomassie Brilliant Blue G-250 in 50 ml of 95% ethanol. Add 100 ml of 85% phosphoric acid, and bring to a final volume of 1 liter with distilled water. Phosphoric acid is extremely corrosive. Handle with care.

Bromophenol Blue (Sodium salt, MW 692.0)

0.01 (w/v) : Dissolve 1 mg of Bromophenol blue, sodium salt to a final volume of 100 ml with either water or buffer.

Calcium Chloride (MW 110.99)

0.0033 M : Dissolve 0.522 grams of calcium acetate to a final volume of 1 liter with water or buffer.

cAMP (Adenosine Monophosphate, cyclic MW 329.22)

0.001M (1mM) : Dissolve 33 mg of cAMP to a final volume of 100 ml with water, buffer, or media.

Coomassie Blue (Coomassie Brilliant Blue R250)

0.25% (w/v) 0.001 M : Dissolve 2.50 grams of Coomassie Brilliant Blue R250 to a final volume of 1 liter with 20% (w/v) trichloroacetic acid (TCA). Some Investigators use a 0.25% solution of Coomassie Blue in methanol-water-glacial acetic acid.

Copper Sulfate ($CuSO_45H_2O$, MW 249.68)

0.5% (w/v) : Dissolve 0.5 grams of copper sulfate to a final volume of 100 ml with water.

Crystal Violet

Dissolve 0.1 grams of crystal violet and 0.25 ml of glacial acetic acid to a final volume of 100 ml with water.

Dinitrophenol (DNP, MW 184.11)

18.4 mg % : Dissolve 18.4 mg of 2, 4-dinitrophenol to a final volume of 100 ml with water or buffer.

Dithiothreitol (Cleland's Reagent, MW 154.3)

0.01 M : Dissolve 154 mg of dithiothreitol to a final volume of 100 ml with water or buffer.

EDTA (Ethylenediaminetetraacetic Acid, MW 292.24)

1 M : Dissolve 292.24 grams of EDTA, free acid to a final volume of 1 liter. If the more soluble disodium salt of EDTA is used, adjust the weight accordingly. The pH can be adjusted with acetic acid or NaOH. For corresponding concentration dilutions, multiply the weight in grams by the desired molarity. For example, for 10 mM EDTA, multiply 292.24 X 0.010 to obtain 2.92 grams of EDTA per liter.

Eosin

0.5% (w/v) : Dissolve 0.5 grams of Eosin Y in 100 ml of water.

Ethanol (C_2H_5OH, MW 46.07)

Density = 0.7893 gm/ml.

50%–95% (v/v) : Since 95% ethyl alcohol is less expensive and easier to store than absolute, these dilutions should be made with 95% ethyl alcohol. Unless otherwise stated, denatured alcohol works as well as the more expensive non-denatured. A simple way to make the % solution is to use the appropriate amount of 95% ethanol and dilute to 950 ml instead of 1 liter. For example, to make a 50% (v/v) solution, measure out 500 ml of 95% ethyl alcohol and dilute to a final volume of 950 ml with water. For a 70% solution, measure 700 ml of ethyl alcohol and dilute to 950 ml with water. Absolute ethanol should be used directly as 100% ethanol. It is important for histology that this be truly 100%. Since it is hydroscopic (it absorbs water from the air), do not assume it is absolute unless it is sealed or treated to ensure no water. To test, add a drop to a sample of xylol. If any cloudiness occurs, the alcohol is not absolute.

8.5 M : The ethanol can be weighed (391.6 grams of absolute, 412.2 grams of 95% (v/v) or measured volumetrically by using the density. That is, 391.6 grams + 0.7893 grams/ml or 496.1 ml of absolute ethanol. Using 95%, 412.2 grams + 0.7893 grams/ml or 522.2 ml. Weigh or measure the appropriate amount and dilute to a final volume of 1 liter with water.

Giemsa Stain

Prepare a stock solution by dissolving 3.8 grams of Giemsa powder in 25 ml of glycerin. Heat gently with stirring for about 2 hours at 60°C. Cool and add 75 ml of methanol (neutral, acetone-free).

For a working solution, dilute the stock solution 1/10 with water before use. For chromosome banding, combine 5.0 ml of stock Giemsa, 3.0 ml absolute methanol, 3.0 ml of 0.1 M citric acid, and 89 ml of distilled water. Adjust the pH of the solution to 6.6 with Na_2HPO_4.

Glycerol (MW 92.09)

10% (v/v) : To 10 m of glycerol (glycerine) add enough water to make a final volume of 100 ml.

8 M : Weigh 73.67 grams of glycerol and add to a final volume of 100 ml. alternatively, measure 499.1 ml of glycerol and make up to a final volume of 1 liter (the density of glycerol at room temperature is 1.476) with water or buffer. For 8 M glycerol in MT buffer, make a 2X MT buffer for use as the diluents.

Glycine (MW 75.07)

0.192 M : Dissolve 1.44 grams of glycine to a final volume of 100 ml with water or buffer.

Hydrochloric Acid (HCl, MW 36.46)

Concentrated HCl has a molarity of approximately 11.6. HCl is a gas, which is soluble in water and comes in the form of concentrated reagent grade HCl. This solution is approximately 36% to 38% (w/v) HCl. To make a 1 N solution, add 86 ml of concentrated HCl to 800 ml of water and dilute to a final volume of 1 liter. For 0.1N, dilute the 1 N by a factor of 10. For % solutions, note that liquid HCl is only 38% HCl, thus a 1% solution would require 2.6 ml of concentrated HCl (1/.38) per final volume of 100 ml.

β-mercaptoethanol (MW 78.13)

00.5 M : Density = 1.2 grams/ml. Use either 3.91 grams OR 3.26 ml of mercaptoethanol in a final volume of 100 ml of water or buffer.

5% (w/v) : Use 5.0 grams or 4.167 ml in a final volume of 100 ml of water or buffer

Phosphate Buffered Saline (PBS)

Mix 100 ml Ca^{++}/Mg^{++} free 10X PBSA with 800 ml of distilled water. Separately, dissolve 0.1 gram of magnesium chloride and 0.1 gram of anhydrous calcium chloride to a final volume of 100 ml with water. With constant stirring, slowly add the magnesium/calcium chloride solution to the diluted PBSA. If a precipitate forms, start over, and add slower, with continuous stirring.

Ca^{++}/Mg^{++} free Phosphate Buffered Saline-10X (10X PBSA)

Dissolve 80 grams of NaCl, 2.0 grams of KCl, 15.0 grams of dibasic sodium phosphate and 2.0 grams of monobasic potassium phosphate in 1 liter of distilled water. This makes a 10X solution of Ca^{++}/Mg^{++} free phosphate buffered saline. Dilute 1:10 prior to use. Store in a refrigerator.

Phosphate-buffered Saline-Tween 20 (4.8)

Mix PBS and add 0.1% (v/v) Tween 20.

Potassium Chloride (KCl, MW 74.55)

1 M : Dissolve 74.55 grams of KCl to a final volume of 1 liter with water or buffer. For other concentrations, multiply the weight by the required molarity. For example, for 0.150 M (150 mM), use 0.150 X 74.55, or 11.183 grams of KCl in 1 liter of water or buffer. Use half as much to obtain 0.075 M for karyotyping.

Potassium Phosphate, Monobasic (KH_2PO_4, MW 136.09)

0.01M : Dissolve 1.36 grams of monobasic potassium phosphate to a final volume of 1 liter with water.

Potassium phosphate, dibasic (K_2HPO_4, MW 174)

0.01M : Dissolve 1.74 grams of dibasic potassium phosphate to a final volume of 1 liter with water.

Potassium phosphate buffer

0.01M, pH 7.4 : Prepare 500 ml of 0.01 M K_2HPO_4 and 500 ml of 0.01 M KH_2PO_4. Place the K_2HPO_4 onto a magnetic stirrer and insert a pH electrode. Add the KH_2PO_4 slowly to adjust the pH to 7.4.

Potassium hydroxide (KOH, MW 56.10)

0.5 N : Dissolve 28.05 grams of KOH to a final volume of 1 liter with water. 10% (w/v): Dissolve 10 grams of KOH to a final volume of 100 ml with water. Store in a plastic container.

Ribonuclease

0.1% (w/v) : Dissolve 10 mg of pancreatic ribonuclease type A in 10 ml of water or buffer. Use for enzyme treatment of histological sections by floating 0.5 to 1.0 ml of this solution onto the section, with the slide set into a covered petro plate.

Safranin

Dissolve 2.5 grams of Safranin O in 10 ml of 95% ethanol and dilute to 100 ml with water.

Saline (NaCl)

0.85% (w/v) : Saline refers to a solution of NaCl, with the most common usage for that which is isotonic to mammalian blood cells, notably a 0.85% or 0.9% solution. To mix, dissolve 8.5 grams of NaCl to a final volume of 1 liter with water.

Silver nitrate solution (for electrophoresis staining)

Dissolve 0.15 grams of NaOH in 150 ml of water. Add 3.5 ml of concentrated NH_4OH and bring to a volume of 200 ml. Separately, dissolve 2.0 grams of silver nitrate in a final volume of 10 ml. With constant stirring, add 8.0 ml of the silver nitrate to the 200 ml of $NaOH/NH_4OH$.

This solution should be prepared immediately prior to use, and used within

30 minutes. Dispose of this solution with copious flushing. It becomes explosive upon drying.

Sodium acetate (MW 82.04)

1 M : Dissolve 82.04 grams of sodium acetate to a final volume of 1 liter with water or buffer.

0.02 M : Dissolve 1.64 grams of sodium acetate to a final volume of 1 liter with water or buffer.

Sodium acetate buffer

1 M pH 5.7 : To 925 ml of 1 M sodium acetate, add 75 ml of 1 M acetic acid.

Sodium chloride (MW 58.44)

M : For a molar solution of sodium chloride, dissolve 58.44 grams of NaCl to a final volume of 1 liter with water or buffer. For corresponding dilutions, multiply the weight by the molarity required. For example, for 0.05 M, multiply 58.44 by 0.05 or 2.92 grams/liter.

% : For % solutions, they are invariably w/v. For a 1% (w/v) solution, dissolve 1.0 gram of NaCl to a final volume of 100 ml with water or buffer. Multiply the weight by a corresponding change in % for other concentrations.

200, 300, and 400 mOsM: Osmoles for NaCl are calculated as twice the molar concentration. Thus, a 200 mOsM solution would be .100 M NaCl. Likewise, 300 mOsM would be .150 M and 400 mOsM would be .200 M NaCl.

Sodium lauryl sulfate (SDS or SLS, MW 288.38)

0.1% (w/v) : Dissolve 0.1 grams of SDS to a final volume of 100 ml with water or buffer. Mix by gentle stirring, do not shake.

10% (w/v) : Dissolve 10 grams of SDS to a final volume of 100 ml with water.

SDS should not be inhaled in its powder form. When weighing, use a mask, or better, a hood.

Sulfuric acid (H_2SO_4, MW 98.08)

Caution : Sulfuric acid is extremely caustic and will cause severe burns. It must always be added to the water, when making dilutions. Upon addition to water or alcohol, heat will be generated while the solution will contract in volume. Use extreme care in handling this acid. Concentrated H_2SO_4 is 17.8 M or 35.6 N. Add 30.6 ml of concentrated sulfuric acid slowly, with constant stirring, and adequate protection from splashes, to approximately 800 ml of water. Cool and make up volume to 1 liter with water.

TEMED (N,N,N,N'-tetramethylethelenediamine)

Catalyst for PAGE. Use directly and add 10 ml TEMED per 15 ml of gel solution.

Trichloroacetic acid (TCA) (CCl_3COOH, MW 163.4)

This is extremely caustic acid. Handle with care. 72% (w/v). A Dissolve 72 gram of TCA to a final volume of 100 ml. TCA is hydroscopic and will readily absorb water. The solid crystals will become liquid if the stock bottle is placed in warm water, with a loose cap (melting point 57°C to 58°C. It is easier to handle as a liquid. Storage of solutions greater than 30% (w/v) are not recommended, as decomposition is rapid. Therefore, these solutions should be made as needed.

Tris buffer

There are many variations on the basic Tris-HCl buffer combination, most of which are commercially available. Solutions with EDTA are known as TE (Tris-EDTA) buffers, while solutions with EDTA and acetic acid are known as TAE buffers. The terminology varies with the author, with Tris buffer being used to mean Tris- HCl solutions. The company, Sigma Chemical Co., St. Louis, carries a full line of the buffers marketed under the trade name of Trizma (base and HCl). The basic buffer is a combination of Tris (Tris (hydroxymethyl) amino methane) and HCl acid. These are sometimes referred to as Tris-base and Tris-HCl solutions. Tris buffers should not be used below a pH of 7.2 or above a pH of 9.0. Tris buffers are also extremely temperature sensitive. Directions are given for room temperature (25°C). The pH will decrease approximately 0.028 units for each degree decrease in temperature.

1M : Dissolve 121 grams of Tris in 800 ml of distilled water. Adjust the pH with concentrated HCl. Dilute to a final volume of 1 liter. Lower required molarities can be diluted from this stock or mixed as combinations of lower molarities of Tris and HCl. It is important to measure the pH at the temperature and molarity that will be used in the final analysis.

Appendix—C

STAINS REQUIRED FOR MICROBIOLOGY EXPERIMENTS

I. Staining Reagents

Acid Alcohol

Ethyl alcohol (95%)	97.0 ml
HCl	3.0 ml

Methylene Blue

Methylene blue	0.3 gm
Distilled water	100.0 ml

Capsule Stain

Crystal violet (1%)	
Crystal violet (85% dye content)	1.0 gm
Distilled water	100.0 ml

Fungal Stains

Lactophenol-cotton blue solution	
Lactic acid	20.0 ml
Phenol	20.0 gm

Glycerol	40.0 ml
Distilled water	20.0 ml
Aniline Blue	0.05 gm

Gram's stain

Crystal violet

Solution A

Crystal violet (90% dye content)	2.0 gm
Ethyl alcohol (95%)	20.0 ml

Solution B

Ammonium oxalate	0.8 gm
Distilled water	80.0 ml

Mix solutions A and B

Gram's iodine

Iodine 1.0 gm

Potassium iodide	2.0 gm
Distilled water	300 ml

Nigrosin

Nigrosin, Water-soluble	10.0 gm
Distilled water	100 ml

Immerse in boiling water bath for 30 minutes.

Formalin	0.5 ml

Filter twice through double-filter paper.

Spore Stain

Malachite green

Malachite green	5.0 gm
Distilled water	100 ml

Acid-fast stain

Carbol fuchsin

Solution A

Basic fuchsin	0.3 gm
Ethyl alcohol (95%)	10.0 ml

Solution B

Phenol	5.0 ml
Distilled water	100 ml

Mix solutions A and B. Add 2 drops of triton-X per 100 ml of stain for use in the heatless method.